DEFENSE, SECURITY AND STRATEGIES

ALTERNATIVE FUEL USE BY THE DEPARTMENT OF DEFENSE

INITIATIVES AND OPPORTUNITIES

DEFENSE, SECURITY AND STRATEGIES

Additional books in this series can be found on Nova's website
under the Series tab.

Additional E-books in this series can be found on Nova's website
under the E-book tab.

DEFENSE, SECURITY AND STRATEGIES

ALTERNATIVE FUEL USE BY THE DEPARTMENT OF DEFENSE

INITIATIVES AND OPPORTUNITIES

RASMUS KLAVAN
EDITOR

nova publishers
New York

For permission to use material from this book please contact us:
Telephone 631-231-7269; Fax 631-231-8175
Web Site: http://www.novapublishers.com

Library of Congress Cataloging-in-Publication Data

ISBN: 978-1-62808-233-3

Published by Nova Science Publishers, Inc. † New York

CONTENTS

PREFACE

This book provides background information and identifies issues for Congress regarding Department of Defense (DOD) alternative fuel initiatives, an issue of considerable attention during hearings in 2012 on DOD's FY2013 budget. Ongoing alternative fuel efforts of the DOD and the military services include purchases of alternative fuels for testing and evaluation, as well as the certification of alternative fuels for use in service fleets. In addition, the Navy, in coordination with the Department of Energy and the Department of Agriculture, intends to spur domestic advanced biofuel production at a commercial scale using the authority of the Defense Production Act. The services (Army, Navy, and Air Force) have spent approximately $48 million on alternative fuels, and the Navy has proposed a $170 million investment in biofuel production capacity. An overview of the DOD's alternative fuels policy and data on DOD's alternative fuels purchase to date, as well as the status of testing platforms on alternative fuel blends and the certification of those blends for fleet-wide use within the services is also discussed.

Chapter 1 – This report provides background information and identifies issues for Congress regarding Department of Defense (DOD) alternative fuel initiatives, a subject of debate at congressional hearings on DOD’s proposed FY2013 budget. The services (the Army, Navy, and Air Force) have spent approximately $48 million to purchase alternative fuels, and the Navy has proposed a $170 million investment in biofuel production capacity. The services have also spent funds on testing, certification and demonstrations of alternative fuels. By comparison, DOD purchases of petroleum fuels totaled approximately $17.3 billion in FY2011.

DOD officials have said that any alternative fuels for DOD operational use must

- be “drop-in;” that is, requiring no modification to existing engines, and
- be cost-competitive with conventional petroleum fuels.

Other desirable characteristics include

- production from a non-food crop feedstock; and
- lifecycle greenhouse gas emissions less than or equal to conventional petroleum fuels.

Each military service has different alternative fuel goals. The Army has the broad aim of increasing the use of renewable energy, but has not adopted any specific alternative fuel goals. The Air Force goals are to test and certify all aircraft and systems on a 50:50 alternative fuel blend by 2012, and to be prepared to acquire 50% of the Air Force's domestic aviation fuel as an alternative fuel blend by 2016. The Navy's goals are to deploy a "Great Green Fleet" strike group of ships and aircraft running entirely on alternative fuel blends by 2016 and to meet 50% of the Navy's total energy consumption from alternative sources by 2020. To meet this goal for its ships, the Navy would need to replace approximately 8 million barrels of petroleum used in its ships with unblended alternative fuels by 2020.

The Navy has also entered into a Memorandum of Understanding (MOU) with the Department of Energy and the Department of Agriculture to promote the development of a domestic advanced biofuel industry through the construction of domestic biofuel plants and refineries. Under the Defense Production Act, the Navy and the Department of Energy plan to fund this initiative with $340 million in federal funds for capital investment and production, with at least equal cost-sharing from industry. The Department of Agriculture intends to provide an additional $171 million through the Commodity Credit Corporation to support biofuel feedstocks.

Legislative debate in 2012 related to DOD's alternative fuels efforts has focused on two areas: (1) proposals in the National Defense Authorization Act for FY2013 (H.R. 4310, S. 3254) to maintain or limit DOD's ability to purchase alternative fuels and invest in biofuel production capability, and (2) appropriations related to the joint Navy, Department of Energy, and Department of Agriculture biofuel production initiative.

Additional areas for potential congressional oversight include the costs and benefits to DOD of alternative fuels, as well as the coordination of alternative fuel initiatives within the services and between DOD and other federal agencies.

Chapter 2 – The Secretaries of Energy, Agriculture, and the Navy have entered into a Memorandum of Understanding (MOU) to "assist the development and support of a sustainable commercial biofuels industry." The objective of the MOU is the construction or retrofitting of multiple domestic commercial or pre-commercial scale advanced drop-in biofuel plants and refineries. The MOU would support the Navy's goal of deploying a "Green Strike Group" by the end of 2012, and "Great Green Fleet" by 2016 fueled in part with a 50/50 blend of hydrotreated renewable jet fuel (biofuel). The Navy proposes to use authority under the Defense Production Act of 1950 (DPA) to develop a domestic industrial capacity to supply biofuel. In its FY2013 Congressional Budget Request, the Department of Energy (DOE) requested authority to transfer funds to the DPA Fund, offering the justification that it will support the MOU with the technical expertise to move pilot-scale demonstration projects to larger-scale production in support of the Navy's Green Fleet Goal. Agriculture, Energy, and the Navy expect to fund this initiative at $510 million in aggregate over three years.

In the past, Congress has found it in the interest of national defense preparedness for government to assure that a domestic industrial capacity exists to produce fuel. Congress set aside the (now depleted) Naval Oil Reserves and Oil Shale Reserves to provide for the Navy's fuel requirements. Congress later promoted alternative fuel from coal through the U.S. Synthetic Liquid Fuels Act of 1944 to aid the execution of World War II, and to conserve and increase national oil resources. The act authorized the Secretary of the Interior to construct, maintain, and operate plants producing synthetic liquid fuel from coal, oil shale, and agricultural and forestry products. During the Korean War, the DPA authorized the

President to have liquid fuels processed and refined for government use or resale, and to make improvements to government- or privately-owned facilities engaged in processing and refining liquid fuels when it would aid the national defense. In 1980, Congress amended the DPA to authorize the President's purchase of synthetic fuels for national defense. Most recently, the Energy Policy Act of 2005 directed the Secretary of Energy, in cooperation with the Secretaries of the Interior and Defense, to develop a program to accelerate the commercial development of strategic unconventional fuels, including but not limited to oil shale and tar sands resources within the United States. Except for exploiting the Naval Oil Reserve, policies that directed alternative fuel development for national defense interests have had to challenge newly discovered petroleum resources that presented clear economic advantages over alternative fuels.

Domestic crude oil production in the United States has increased over the past few years, reversing a decline that began in 1986. The United States is now a net exporter of refined petroleum products. Over the next 10 years, continued development of unconventional oil resources, in combination with the ongoing development of offshore resources in the Gulf of Mexico may push domestic crude oil production to a level not seen since 1994, according to the U.S. Energy Information Administration.

An important policy question for Congress may be whether a domestic biofuel industry is necessary for national defense, and whether proceeding under the authority of the DPA offers the necessary stimulus. A domestic biofuel industry may satisfy concerns for a secure, domestic, alternative fuel source independent of unstable foreign petroleum suppliers. However, adding biofuel to the military's supply chain does not relieve logistical issues with delivering fuel to forward operating areas, where fuel supply issues have been more about vulnerability than availability.

Chapter 3 – Section 334 of the National Defense Authorization Act (NDAA) for Fiscal Year 2010 requires the Department of Defense (DoD) to provide an assessment of the use of renewable fuels in non-tactical and tactical aviation, maritime, and ground transportation fleets and asks whether establishing a DoD commodity class for renewable fuels distinct from petroleum-based products would be beneficial. The Office of the Assistant Secretary of Defense for Operational Energy Plans and Programs and the Defense Logistics Agency, in conjunction with the military Services, prepared this report.

This report was prepared in response to the Section 334 requirement and assesses renewable fuel supply (anticipated feedstock availability, production capacity and production) and demand (projected fuel quantities based on the military Services' requirements and plans) through 2020. To do so, the report reviews statutory, regulatory, and other drivers relevant to DoD use of alternative and renewable fuels; discusses relevant DoD and Service policies, programs, and goals; projects the rapidly changing U.S. domestic renewable fuels market; and examines the implications for DoD renewable fuel use by 2020.

In: Alternative Fuel Use by the Department of Defense
Editor: Rasmus Klavan
ISBN: 978-1-62808-233-3

Chapter 1

DOD ALTERNATIVE FUELS: POLICY, INITIATIVES AND LEGISLATIVE ACTIVITY*

Katherine Blakeley

SUMMARY

This report provides background information and identifies issues for Congress regarding Department of Defense (DOD) alternative fuel initiatives, a subject of debate at congressional hearings on DOD's proposed FY2013 budget. The services (the Army, Navy, and Air Force) have spent approximately $48 million to purchase alternative fuels, and the Navy has proposed a $170 million investment in biofuel production capacity. The services have also spent funds on testing, certification and demonstrations of alternative fuels. By comparison, DOD purchases of petroleum fuels totaled approximately $17.3 billion in FY2011.

DOD officials have said that any alternative fuels for DOD operational use must

- be "drop-in;" that is, requiring no modification to existing engines, and
- be cost-competitive with conventional petroleum fuels.

Other desirable characteristics include

- production from a non-food crop feedstock; and
- lifecycle greenhouse gas emissions less than or equal to conventional petroleum fuels.

Each military service has different alternative fuel goals. The Army has the broad aim of increasing the use of renewable energy, but has not adopted any specific alternative fuel goals. The Air Force goals are to test and certify all aircraft and systems on a 50:50 alternative fuel blend by 2012, and to be prepared to acquire 50% of the Air Force's domestic aviation fuel as an alternative fuel blend by 2016. The Navy's goals are to deploy a "Great Green Fleet" strike group of ships and aircraft running entirely on alternative fuel blends by 2016 and to meet 50% of the Navy's total energy consumption

* This is an edited, reformatted and augmented version of Congressional Research Service, Publication No. R42859, dated April 11, 2013.

from alternative sources by 2020. To meet this goal for its ships, the Navy would need to replace approximately 8 million barrels of petroleum used in its ships with unblended alternative fuels by 2020.

The Navy has also entered into a Memorandum of Understanding (MOU) with the Department of Energy and the Department of Agriculture to promote the development of a domestic advanced biofuel industry through the construction of domestic biofuel plants and refineries. Under the Defense Production Act, the Navy and the Department of Energy plan to fund this initiative with $340 million in federal funds for capital investment and production, with at least equal cost-sharing from industry. The Department of Agriculture intends to provide an additional $171 million through the Commodity Credit Corporation to support biofuel feedstocks.

Legislative debate in 2012 related to DOD's alternative fuels efforts has focused on two areas: (1) proposals in the National Defense Authorization Act for FY2013 (H.R. 4310, S. 3254) to maintain or limit DOD's ability to purchase alternative fuels and invest in biofuel production capability, and (2) appropriations related to the joint Navy, Department of Energy, and Department of Agriculture biofuel production initiative.

Additional areas for potential congressional oversight include the costs and benefits to DOD of alternative fuels, as well as the coordination of alternative fuel initiatives within the services and between DOD and other federal agencies.

INTRODUCTION

This report provides background information and identifies issues for Congress regarding Department of Defense (DOD) alternative fuel initiatives, an issue of considerable attention during hearings in 2012 on DOD's FY2013 budget.

Ongoing alternative fuel efforts of the DOD and the military services include purchases of alternative fuels for testing and evaluation, as well as the certification of alternative fuels for use in service fleets. In addition, the Navy, in coordination with the Department of Energy and the Department of Agriculture, intends to spur domestic advanced biofuel production at a commercial scale using the authority of the Defense Production Act.

The services (Army, Navy, and Air Force) have spent approximately $48 million on alternative fuels, and the Navy has proposed a $170 million investment in biofuel production capacity. The services have also spent funds on testing, certification and demonstrations of alternative fuels. By comparison, DOD purchases of petroleum fuels totaled approximately $17.3 billion in FY2011.

This report provides a brief overview of DOD alternative fuels policy and data on DOD's alternative fuels purchases to date, as well as the status of testing platforms on alternative fuel blends and the certification of those blends for fleet-wide use within the services. This report also discusses the current status of the Navy's biofuel production initiative under the Defense Production Act, in conjunction with the Department of Energy and Agriculture, including appropriated funding. It also provides the status of recent legislative actions related to DOD alternative fuels efforts.

DOD Alternative Fuels Policy

Section 314 of the FY2012 National Defense Authorization Act[1] gave responsibility for and oversight of DOD's alternative fuels initiatives and policy to the Assistant Secretary of Defense for Operational Energy Plans and Programs.[2] Previously, there was no specific responsibility or oversight of alternative fuel policy and initiatives at the DOD-wide level. The statute requires that the Assistant Secretary of Defense for Operational Plans and Programs

- lead DOD's alternative fuel activities;
- oversee DOD's alternative fuel investments;
- make recommendations regarding the development of alternative fuels by the military departments and the Office of the Secretary of Defense;
- issue guidelines and prescribe policy to streamline alternative fuels investments across DOD; and
- encourage collaboration and leverage the investments in alternative fuel development made by the Department of Energy, the Department of Agriculture, and other federal agencies to the benefit of DOD.

Policy Goals

On July 5, 2012, Assistant Secretary of Defense for Operational Energy Plans and Programs Sharon Burke issued a memorandum articulating the goals of DOD alternative fuels policy, which are to "ensure operational military readiness, improve battlespace effectiveness," and increase "the ability to use multiple, reliable fuel sources." Specific considerations for DOD investments in alternative fuels initiatives include "increased resilience against strategic supply disruptions; dampened effect[s] of petroleum price volatility, increased fuel options for operational commanders and ultimately increased expeditionary effectiveness."[3]

Pursuant to these goals, DOD officials have said that any alternative fuels for DOD operational use must:

- be "drop-in," that is, requiring no modification to existing engines;
- be cost-competitive with conventional petroleum fuels;
- be available in sufficient quantities.

Other desirable characteristics include

- production from a non-food crop feedstock; and
- lifecycle greenhouse gas emissions less than or equal to conventional petroleum fuels.[4]

Investments in Alternative Fuels

By statute,[5] DOD investments in alternative fuel activities must be certified as part of the annual operational energy budget certification process. The Assistant Secretary of Defense for Operational Energy Plans and Programs must review the services' proposed budgets, and certify whether these budgets are adequate to implement the operational energy strategy.[6] The annual operational energy report[7] must now incorporate alternative fuels initiatives, including descriptions, funding, and expenditures. Per the July 5, 2012, memorandum, future investments in alternative fuels will be subject to a "rigorous, merit-based evaluation." DOD alternative fuels development initiatives will generally follow three phases:

Phase 1: Certification: A cross-services fuels working group is required to submit an annual plan that identifies promising alternative fuels and coordinates certification efforts by the services.

Phase 2: Field Demonstration: The services may propose a field demonstration of a new fuel. Depending on the funding source the proposal will be reviewed by the Office of the Assistant Secretary of Defense for Operational Energy Plans and Programs during the annual operational energy budget certification process, or jointly by that office and the Office of the Undersecretary of Defense (Comptroller) and the Office of the Assistant Secretary of Defense (Logistics and Materiel Readiness).

Phase 3: Ongoing Purchases: For purchases of operational quantities of fuels, alternative fuels will compete with petroleum products, and awards will be based on "the ability to meet requirements at best value to the government, including cost."[8]

Service Goals for Use of Alternative Fuels

Each of the military services has set goals regarding alternative fuels.

The Army has not adopted any specific alternative fuel goals. However, the Army does have the broad goal of increasing the use of renewable/alternative energy, set out in the Army Energy Security Implementation Strategy. The Army's Tactical Fuel and Energy Implementation Plan study, released in 2010, recommended the following goals in order to meet this aim of increased renewable/alternative energy: by 2028, at least 25% of energy used for tactical level power generation derived from alternative and renewable sources, and by 2028, 50% of the fuel requirement in the training base for the tactical mobility fleet (surface and air) met by alternative fuel blends.[9]

The Air Force has set a goal of being prepared to "cost-competitively acquire 50% of the Air Force's domestic aviation fuel requirements via an alternative fuel blend in which the alternative component is derived from domestic sources produced in a manner that is greener than fuels produced from conventional petroleum" by 2016.[10] In order to be prepared to use alternative fuels, should they become cost competitive, the Air Force has an additional goal of testing and certifying all aircraft and systems on a 50:50 alternative fuel blend[11] by 2012.[12] (See the Air Force portion of the section "Alternative Fuels Testing and Evaluation" for more details.)

The Navy has set more ambitious goals related to alternative fuels, as part of the Navy's five energy goals.[13] One goal is to deploy a "Great Green Fleet" strike group of ships and aircraft running entirely on alternative fuel blends by 2016. This "Great Green Fleet," demonstrated during the July 2012 RIMPAC naval exercises, is a carrier strike group composed of nuclear-powered ships, ships running on a biofuel blend, and aircraft flying on a biofuel blend.

A second Navy goal is meeting 50% of the Navy's total energy consumption from alternative sources by 2020. To meet this goal for the Navy's use of energy afloat, the Navy plans to reduce its liquid fuel usage afloat to 38 million barrels of oil equivalent per year.

Nuclear power is projected to provide 11 million barrels of oil equivalent in 2020. Therefore, to meet the goal of 50% alternative energy (18 million barrels of oil equivalent in 2020), approximately 8 million barrels of petroleum used afloat would need to be replaced by unblended alternative fuels by 2020.

DOD ALTERNATIVE FUEL PURCHASES

Since 2007, DLA Energy has procured approximately 1.9 million gallons of various types of alternative fuels on behalf of the Army, Navy, and Air Force using funds provided by the services. DOD purchases of alternative fuels in order to test engine performance and certify alternative fuels for use in service fleets have totaled about $48 million to date. **Table 1** provides an overview of each service's alternative fuels purchases from 2007 through the present.

Table 1. Alternative Fuels Purchases by Service, 2007-November 2012

Service	Total Gallons Purchased	Total Cost	Average Cost Per Gallon
Army	49,950	$1,632,120	$ 37.14
Navy	676,500	$20,618,450	$ 30.15
Air Force	1,166,000	$25,719,650	$ 22.06
Total	1,888,450	$47,970,220	$ 25.43

Source: Data from DLA-E, November 9, 2012.

Notes: The average cost per gallon was calculated by dividing the total spending of each service on alternative fuels by number of gallons purchased. Values are not adjusted for inflation.

As of November 26, 2012, the Navy has not exercised its two options to purchase an additional 50,000 gallons of DSH fuel at a cost of $25.73 per gallon from Amyris, for a total cost of $1,286,500.

The services have purchased various types of alternative fuels, including Fischer-Tropsch fuels[14] derived from coal and natural gas, and three types of biofuels (hydrotreated jet and diesel biofuels,[15] alcohol-to-jet fuels, and direct-sugar-to-hydrocarbons fuels).[16] All services have purchased various types of hydrotreated jet and diesel biofuels. The Air Force has also purchased alcohol-to-jet fuels derived from both petroleum and biomass feedstocks. The Navy has purchased some direct-sugar-to-hydrocarbons fuel, derived from bio-based feedstocks. The Air Force has also purchased fuels created via the Fischer-Tropsch process from coal and natural gas. **Table 2** provides an overview of DOD alternative fuel purchases by fuel type.

Section 526 of the Energy Independence and Security Act of 2007[17] prohibits federal agencies, including DOD, from entering into contracts for alternative fuels that have lifecycle greenhouse gas emissions greater than or equal to the emissions from the equivalent conventional petroleum fuel, except for research or testing purposes. This provision would exclude coal-to-liquid fuels, which have lifecycle greenhouse gas emissions of approximately 2.5 times those of petroleum fuels, from purchases of alternative fuels. DOD's position is that EISA section 526 does not apply to fuel purchases where the source of the fuel is not specified in the contract, and therefore does not apply to DLA-E's purchases of bulk fuel in operational quantities.[18]

An overview of DOD's alternative fuel purchases to date is in **Table A-1** in Appendix A.

Alternative Fuels Testing and Evaluation

As part of their alternative fuels initiatives, the Army, Navy, and Air Force have been testing various alternative fuel blends in their equipment, for the potential certification of alternative fuels for fleet-wide use.

Army

The Army is currently testing 50:50 blends of Fischer-Tropsch synthetic paraffinic kerosene and hydrotreated renewable jet with JP-8 for use in all Army ground systems and field generators, with the goal of certifying these fuels by 2014.[19] The Army is also working to obtain Air Force certification for H60 Black Hawk helicopters to fly on the 50:50 FT-SPK: JP-8 blend.

As part of the Army's 2009 Army Energy Security Implementation Strategy, the Army intends to complete testing of tactical ground equipment systems for which alternative or renewable fuels and synthetic fuel blend evaluations are completed by the end of FY2014. For Army engine and aviation systems for which alternative or renewable fuels and synthetic fuel blend evaluations are completed, the Army intends to complete 50% of testing by the end of FY2014 and 100% by the end of FY2016.[20]

Table 2. Alternative Fuel Purchases by Fuel Type, 2007-November 2012

Fuel Type	Gallons	Average cost per gallon	Total Spending
Hydrotreated Renewable Jet (HRJ)/Hydrotreated Renewable Diesel (HRD) both Hydroprocessed Esters and Fatty Acids (HEFA) fuels	1,085,450	$38.26	$41,534,620
Fischer-Tropsch (FT)	730,000	$3.76	$2,745,650
Alcohol-to-Jet (ATJ)	56,000	$59.00	$3,304,000
Direct Sugar to Hydrocarbon (DSH)	15,000	$25.73	$385,950
Total	1,606,450		$47,970,220

Source: DLA-E, November 9, 2012.

Notes: It is difficult to compare the costs per gallon of various purchases, even within the same type of fuel, as purchases may use different feedstocks or production pathways. Additionally, the contracts may include R&D costs in addition to the production costs of the fuel.

Air Force

The Air Force has certified Fischer-Tropsch fuel blends for use in its manned and unmanned fleets. Testing has been completed on biofuels, such as the HEFA fuels described above, and fleet-wide certification is expected shortly. The Air Force has also begun testing alcohol-to-jet fuels, with the inaugural flight of an A-10 Thunderbolt II jet on an ATJ-conventional fuel blend on June 28, 2012. According to the Air Force, this test flight was the first flight powered by an ATJ-conventional fuel blend.[21]

Navy

In April 2010, the "Green Hornet" F/A-18 was the first Navy jet to fly on a biofuel blend. The Navy has tested HEFA 50:50 biofuel blends (also known as hydrotreated renewable jet [HRJ]) for use in all manned and unmanned aircraft. Testing of 50:50 HEFA fuel blends for ships (also known as HRD, or hydrotreated renewable diesel) is complete except for one type of diesel generator. The Navy anticipates certification for all HEFA fuels in early 2013. Testing and certification of Fischer-Tropsch fuels is expected to be completed by the end of 2013. Testing of other alternative fuels, such as alcohol-to-jet, pyrolysis oils, and direct sugar to hydrocarbon, will continue in the future.[22]

Navy Biofuel Production under the Defense Production Act

In June 2011, the Departments of the Navy, Energy, and Agriculture signed a Memorandum of Understanding (MOU) to "assist the development and support of a sustainable commercial biofuels industry." The MOU argues that because of the current economic environment, start-up risks, and competitive barriers posed by the established crude oil market, without government investment, adequate domestic production capacity of advanced drop-in biofuels will not be achieved in a timely manner.[23]

The MOU calls for the Navy, the Department of Energy, and the Department of Agriculture to support advanced drop-in biofuel plants and refineries to produce advanced biofuels that

- meet military specifications;
- are price competitive with petroleum;
- are at geographically diverse locations with ready market access; and
- have no significant impact on the food supply.[24]

Under the authority of the Defense Production Act, the Navy and the Department of Energy plan to fund this initiative with $340 million in federal funds for capital investment and production, with at least equal cost-sharing from industry. The Department of Agriculture will provide an additional $171 million through the Commodity Credit Corporation to support biofuel feedstocks. Total government funding for this project is anticipated at $510 million over a period of three years, with $170 million each from the Navy, the Department of Energy, and the Department of Agriculture.[25]

Funding Opportunity Announcement

The Funding Opportunity Announcement (FOA-12-15-PKM) for this biofuels production initiative under this MOU was initially released on June 27, 2012, with an announced government funding amount of $210 million.

Awards for biofuels production facilities are planned to occur in two phases:

Phase 1: approximately five awards of up to $6 million each for planning and preliminary designs for biofuel production facilities. Phase 1 awards are expected to be announced in March 2013.

Phase 2: up to three awards of up to $70 million each for construction, commissioning, and performance testing of biofuel production facilities.

Requirements for a successful proposal include

- at least 50% cost share for both Phase 1 and Phase 2;
- domestic production of advanced biofuels, including domestic sourcing of feedstocks;
- compliance with Section 526 of the Energy Security and Independence Act of 2007;[26]
- use of an acceptable renewable feedstock;
- production of a drop-in fuel;
- projected cost-competitiveness of the produced fuel with conventional petroleum-based fuels; and
- a production capacity of at least 10 million gallons a year. The stated FOA funding level of $210 million includes
- $100 million from DOD, funded via the $150 million appropriated for the DPA Fund in the FY2012 Department of Defense Appropriations Act;[27]
- $70 million of requested funding for the Navy's drop-in biofuels production initiative as part of the $89 million request for the DPA fund in the DOD's FY2013 budget request;[28] and
- $40 million from DOE following the receipt of the authority, requested in DOE's FY2013 budget request,[29] to transfer monies from DOE Energy Efficiency and Renewable Energy appropriated funds into the DPA Fund to support the MOU between the Navy, DOE, and USDA.

It does not include $171 million of USDA funding via the Commodity Credit Corporation to subsidize the production of bio-based jet fuel,[30] under the broader authority granted to USDA and the Commodity Credit Corporation to increase the use of agricultural commodities under the Commodity Credit Corporation Charter Act.[31]

The Corporation has the authority to borrow up to $30 billion directly from the Treasury or from private lenders, to be repaid, with interest, through appropriations from Congress. This borrowing authority does not require specific congressional appropriations.

Table 3. Funding Sources for the Navy, DOE and USDA Biofuel Production Initiative

		Defense Production Act Fund				**Commodity Credit Corporation**	**Total**
	FY2012	**FY2013**	**FY2014**	**FY2015**	**Subtotal**		
DOD/Navy	$100m	$70m (requested)	—	—	$170m	—	$170m
Department of Energy	—	$40m (requested)	total of $130m across FY2014 and FY2015 (to be requested)		$170m	—	$170m
Department of Agriculture	—	—	—	—	—	$171m	$171m
Total Funding to date:		**$210m** (anticipated, as per the Funding Opportunity Announcement)					**$511m** (anticipated)

Source: Navy energy officials, October 2012.

Notes: DOD and DOE funding requests and amounts for FY2013 can be found in their respective annual budget requests and appropriations bills. The Commodity Credit Corporation has the authority to borrow up to $30 billion directly from the Treasury or from private lenders, to be repaid, with interest, through appropriations from Congress. This borrowing authority does not require specific congressional appropriations. According to Navy officials, the appropriations requests for the balance of the DOE's $170 million funding share after FY2013, $130 million, will be divided between FY2014 and FY2015 requests. The amounts to be requested in each year are not yet known.

LEGISLATIVE ACTIVITY IN 2012

Legislative activity in 2012 related to DOD's alternative fuels efforts has focused on two areas: (1) proposals to expand or limit DOD's ability to purchase alternative fuels and invest in alternative fuel production capability, and (2) appropriations related to the Navy's biofuel production efforts under the DPA.

Restrictions on DOD's Purchase or Production of Biofuels

H.R. 4310 as passed by the House on May 28, 2012, contained two provisions (Section 313 and Section 314) that would exempt DOD from Section 526 of EISA (which requires all alternative fuels purchased by the federal government for operational use to have lifecycle greenhouse gas emissions less than or equal to those from conventional fuels) and limit its ability to purchase alternative fuels that are more expensive than comparable petroleum fuels, respectively.

S. 3254, as passed by the Senate on December 4, 2012, did not contain any provisions that would restrict DOD's abilities to purchase alternative fuels or invest in biofuels production capability. Two restrictive provisions added via amendments during the Senate Armed Services Committee markup (Section 314 and Section 2823, as reported June 4, 2012) were removed by S.Amdt. 2985 and S.Amdt. 3095, respectively, during floor debate.

Section 315 of the enrolled FY2013 National Defense Authorization Act (H.R. 4310/P.L. 112-239) states:

> Amounts made available to the Department of Defense pursuant to the Defense Production Act of 1950 (50 U.S.C. App. 2061 et seq.) for fiscal year 2013 for biofuels production may not be obligated or expended for the construction of a biofuel refinery until the Department of Defense receives matching contributions from the Department of Energy and equivalent contributions from the Department of Agriculture for the same purpose.

DOD and DOE Appropriations for Biofuels Production Initiative

The Departments of Defense and Energy have also requested $70 million and $40 million in appropriations for FY2013 to fund the joint Departments of Defense, Energy, and Agriculture biofuels production initiative under the Defense Production Act.

Department of Defense Appropriations

The Navy and the Department of Energy shares of the MOU with the Department of Agriculture to assist the development of a commercial-scale domestic advanced biofuel industry rely on appropriated funds.

The USDA portion of the funds would be provided through the Commodity Credit Corporation.

For FY2013, the Navy requested $70 million for advanced drop-in biofuels production, as part of the $89 million total requested appropriations to the Defense Production Act Fund.[32]

The House Appropriations Committee report on FY2013 Department of Defense Appropriations Act (H.R. 5856, H.Rept. 112-493) declined to fund the $70 million requested for the joint biofuel production initiative as part of the DPA Fund.

As passed by the House on July 19, 2012, H.R. 5856 contains $63.5 million in DPA funding to remain available until expended. The corresponding Senate Appropriations Committee report (S.Rept. 112-196) recommends funding the overall DPA Fund at $100 million over the amount requested in the budget submission, for a total of $189 million, with no specific mention of biofuel production.

The Department of Defense, Military Construction and Veterans Affairs, and Full-Year Continuing Appropriations Act, 2013 (H.R. 933/P.L. 113-6) includes the FY2013 DOD appropriations as Division C.

This act appropriates $223,531,000 for the Defense Production Act Fund, an increase of about $134.3 million over the amount requested. However, the explanatory statements of both the House and the Senate, as passed by the Senate on March 20, 2013 and the House on March 21, 2013, reduced the funding requested for production of advanced drop-in biofuels by $10 million as "ahead of need," for an appropriation of $60 million.

H.R. 933/P.L. 113-6 also included $20 million in funding for research into lowering the emissions of coal to liquid fuels, in Research & Development, Air Force. The explanatory statements of the Senate (as passed March 20, 2013) and House (as passed March 20, 2013):

> PROMOTING ENERGY SECURITY
>
> The conferees do not include a provision as proposed by the House regarding the Energy Independence and Security Act. However, the conferees provide $20,000,000 in Research, Development, Test and Evaluation, Air Force only for research that will improve emissions of coal to liquid fuel to enable this technology to be a competitive alternative energy resource to meet the goals established in the Department of Defense's Operational Energy Strategy and its Implementation Plan.
>
> The conferees direct the Secretary of the Air Force, in consultation with the Assistant Secretary of Defense for Operational Energy Plans and Programs, to inform the congressional defense committees 30 days prior to any obligation or expenditure of these funds.

Department of Energy Appropriations

For FY2013, DOE requested the authority to transfer $40 million from its Energy Efficiency and Renewable Energy appropriations into the Defense Production Act Fund.

The House-passed version of the FY2013 Energy and Water Development and Related Agencies Appropriations Act (H.R. 5325, as passed on June 6, 2012) does not include any language authorizing the Department of Energy to transfer money into the Defense Production Act Fund, while the Senate version (S. 2465, as reported on April 26, 2012) allows the Department of Energy to transfer up to $100 million of Energy Efficiency and Renewable Energy funds into the Defense Production Act Fund.

Under the Department of Defense, Military Construction and Veterans Affairs, and Full-Year Continuing Appropriations Act, 2013 (H.R. 933/P.L. 113-6), the Department of Energy funding was appropriated under a continuing resolution.

DOE's Energy Efficiency and Renewable Energy was funded at approximately $1.8 billion.

For a more detailed history of recent legislative action, please see Appendix B.

QUESTIONS FOR CONGRESS

DOD and the services' alternative fuels initiatives raise several potential oversight questions and issues for Congress, including the following:

Benefits and Costs of Alternative Fuels

- What benefits for the services do alternative fuels offer over conventional petroleum fuels? Are there alternative ways to achieve these benefits?
- Should these alternative fuel efforts be viewed in terms of their potential benefits to the services, or in terms of their potential benefits to the nation?

- How much funding should be invested in the services' alternative fuels testing and certification efforts?
- To what extent are the services coordinating their alternative fuels testing and certification efforts to prevent duplication of effort?

Navy Role in Developing Advanced Biofuels

- What analysis by the Navy, DOE and USDA supports the Navy's conclusion that this $510 million investment would positively impact the domestic advanced biofuel industry?
- Should biofuel investment be left to private industry (e.g., commercial aviation) or other government agencies such as the Department of Energy or the Department of Agriculture?
- Why is the DOD effort to jumpstart a domestic advanced biofuels industry being led by the Navy?

DOD Coordination of Service Alternative Fuel Initiatives

- How well is the office of the Assistant Secretary of Defense for Operational Energy Plans and Programs (ASD(OEPP))[33] coordinating the alternative fuel goals of the military services?
- How is ASD(OEPP) overseeing current investments by the military services in alternative fuels testing and certification? Is ASD(OEPP)'s oversight authority, including the required budget certification process, adequate?
- Has ASD(OEPP)'s coordination activities or alternative fuels guidance and policy affected the alternative fuel goals and initiatives of the military services? If so, what have been these changes?

Coordination of Alternative Fuel Initiatives between DOD and Other Federal Agencies

- Is there overlap or duplication between DOD's alternative fuel initiatives and the alternative fuel initiatives being pursued by other federal agencies?
- Does the executive branch use a process to coordinate alternative fuel and other energy initiatives across all federal agencies? If so, what are the steps of this process and what criteria are used to determine whether an initiative should be pursued by DOD or some other federal agency?

APPENDIX A. DOD ALTERNATIVE FUEL CONTRACTS TO DATE

Table A-1. DOD Alternative Fuel Contracts, 2007-November 2012

Contract Number	Company	Product	Contract Award Data or Option Date	Gallons	Cost per Gallon	Total Cost	Feedstock	Service	Funding Source
07-D-0486	Shell	FT Iso	6-Jun-07	315,000			Natural Gas	Air Force	Air Force
		Paraffinic Kerosene			$3.41	$1,074,150			RDT&E
08-D-0496	SASOL	FT Iso	26-Jun-08	60,000			Coal	Air Force	Air Force
		Paraffinic Kerosene			$3.75	$225,000			RDT&E
08-D-0497	SASOL	FT Iso	3-Jul-08	335,000			Coal	Air Force	Air Force
		Paraffinic Kerosene			$3.90	$1,306,500			RDT&E
09-D-0519	Sustainable	HRJ5	31-Aug-09	40,000			Camelina	Navy	Navy & DLA
	Oils				$66.60	$2,664,000			ARRA RDT&Ea
09-D-0518	Solazyme	HRJ5	1-Sep-09	1,500	$149.00	$223,500	Algal Oil	Navy	DLA ARRA RDT&E
09-D-0520	Sustainable Oils	HRJ8	15-Sep-09	100,000	$66.80	$6,680,000	Camelina	Air Force	Air Force RDT&E
09-D-0517	UOP	HRJ8	15-Sep-09	100,000	$64.00	$6,400,000	Tallow	Air Force	Air Force RDT&E
09-D-0523	PM Group Int'l	FT F76	30-Sep-09	20,000	$7.00	$140,000	Nat Gas	Navy	Navy RDT&E
10-D-0489	Sustainable Oils	HRJ8	26-Jul-10	34,950	$38.60	$1,349,070	Camelina	Army	DLA ARRA RDT&E
Option	Sustainable	HRJ5	29-Jun-10	150,000			Camelina	Navy	Navy RDT&E;
	Oils				$34.45	$5,167,500			DLA ARRA RDT&E
Option	Sustainable Oils	HRJ8	31-Aug-10	100,000	$34.90	$3,490,000	Camelina	Air Force	Air Force RDT&E
Option	UOP	HRJ8	31-Aug-10	100,000	$32.40	$3,240,000	Tallow	Air Force	Air Force RDT&E
11-D-0526	Gevo	ATJ8	23-Sep-11	7,000	$59.00	$413,000	Alcohols	Air Force	Air Force RDT&E
Option	Gevo	ATJ8	28-Sep-11	4,000	$59.00	$236,000	Alcohols	Air Force	Air Force RDT&E
11-D-0530	UOP	HRJ8	30-Sep-11	4,500	$33.00	$148,500	Camelina	Army	Army RDT&E

Table A-1. (Continued)

Contract Number	Company	Product	Contract Award Data or Option Date	Gallons	Cost per Gallon	Total Cost	Feedstock	Service	Funding Source
12-D-0549	Dynamic	HRD76	30-Nov-11	350,000	$26.75	$9,362,500	UCO/Algal	Navy	Navy OM&N
12-D-0549	Dynamic	HRJ5	30-Nov-11	100,000	$26.75	$2,675,000	UCO/Algal	Navy	Navy OM&N
12-D-0559	UOP	HRJ8	2-May-12	4,500	$29.90	$ 134,550	UCO/ICO	Army	Army RDT&E
12-D-0560	Amyris	DSH76	27-Sep-12	15,000	$25.73	$ 385,950	Ferm. Sugar	Navy	Navy RDT&E
Option	Amyris	DSH76	TBD	25,000	$25.73	$ 643,250	Ferm. Sugar	Navy	Navy RDT&E
Option	Amyris	DSH76	TBD	25,000	$25.73	$ 643,250	Ferm. Sugar	Navy	Navy RDT&E
12-D-0561	Gevo	ATJ8	27-Sep-12	30,000	$59.00	$ 1,770,000	Alcohols	Air Force	Air Force RDT&E
Option	Gevo	ATJ8	27-Sep-12	15,000	$59.00	$ 885,000	Alcohols	Air Force	Air Force RDT&E
TOTAL:				1,886,450		$47,970,220			

Source: Data from DLA-E, November 9, 2012.

Note: As of November 26, 2012, the Navy has not exercised its two options to purchase a total of 50,000 additional gallons of DSH fuel at a cost of $25.73 per gallon from Amyris, for a total cost of $1,286,500. If the Navy exercises these options, DOD purchases of biofuels to date will total 1,936,450 gallons at a total cost of $49,256,720.

a. DLA ARRA RDT&E refers to funds from the American Recovery and Reinvestment Act of 2009 (P.L. 111-5).

APPENDIX B. RECENT LEGISLATIVE ACTION

FY2012 Department of Defense Appropriations Act (P.L. 112-74)

The FY2012 Department of Defense Appropriations Act (combined into the Consolidated Appropriations Act of FY2012 [H.R. 2055/P.L. 112-74]) included $150 million in appropriated funds to support DPA Title III funding. While not specifically directing this additional $150 million of DPA Title III funding to biofuels production, both the House and Senate accompanying reports were supportive of DOD biofuels initiatives and encouraged longer contract terms for biofuels procurement. The Senate Appropriations Committee report (S.Rept. 112-77 of September 15, 2011, accompanying H.R. 2219) stated:

> *Long-term Contracts.*—The Committee is encouraged by the Department's biofuels initiatives such as the Green Fleet program; however, the Committee is concerned that the Department lacks the long-term contracting ability to ensure that adequate biofuels are produced. To address this issue, the Department may fund multi-year contracts with purchase periods up to 15 years for biofuels products in order to maximize efficiencies of scale for the best purchase price. (page 160).

The House Appropriations Committee report (H.Rept. 112-331 of December 15, 2011, accompanying H.R. 2055), stated:

> LONG TERM CONTRACTS
>
> The conferees believe that the time and money being invested by the Department of Defense in biofuels and alternative energy will reap dividends not only for the Nation's armed forces, but eventually for the Nation itself. The conferees want the Department to be in the best position possible to take advantage of the expected breakthroughs in this area and encourage the Department to eventually pursue extended multi-year contracts, pursuant to the Financial Management Regulation, for biofuel products in order to maximize efficiencies of scale for the best purchase price. (p. 671)

FY2013 Department of Energy Appropriations Act (H.R. 5325/S. 2465)

In its FY2013 Congressional Budget Request, the Department of Energy (DOE) requested authority to transfer funds to the DPA Fund, offering the justification that it will support the MOU with the technical expertise to move pilot-scale demonstration projects to larger-scale production.[34]

The House-passed version of the Energy and Water Development and Related Agencies Appropriations Act, FY2013, (H.R. 5325, as passed on June 6, 2012) did not include any language authorizing DOE to transfer money into the Defense Production Act Fund. The House Appropriations Committee, in its accompanying report (H.Rept. 112-462 of May 2, 2012), declined to fund and authorize the Department of Energy's portion of this joint initiative. The report states:

> *Biomass and Biorefinery Systems R&D.*—... The Committee recommends $203,000,000 for Biomass and Biorefinery Systems R&D, $3,000,000 above fiscal year

2012 and $67,000,000 below the budget request. The Department is directed to continue conducting only research, development, and demonstration activities advancing technologies that can produce fuels and electricity from biomass and crops that could not otherwise be used as food. The budget request proposed funding and legislative language for a joint initiative with the Navy and the Department of Agriculture to develop commercial diesel and jet biofuels production capacity for defense purposes. The Department has not adequately justified why the Department of Energy should fund this Defense initiative, and whether the proposed investments can successfully lower costs to competitive levels in several years or will only serve to sink costs into a product that is too immature to compete without federal support. The recommendation includes no funding for the proposed initiative and does not include the requested legislative language. (p. 85-86)

In FY2013, the Senate Appropriations Committee Energy and Water Development and Related Agencies Appropriations Act (S. 2465, as reported on April 26, 2012) provides that, of the funds appropriated for Energy Efficiency and Renewable Energy, "the Secretary may transfer up to $100,000,000 to the Defense Production Act Fund for activities of the Department of Energy pursuant to the Defense Production Act of 1950 (50 U.S.C. App. 2061, et seq.)."

The accompanying report (S.Rept. 112-164 of April 26, 2012) includes language supportive of the DOD, DOE, and USDA joint biofuel production initiative. The report states:

> *Biomass and Biorefinery Systems R&D.*—The Committee recognizes that quality and reliability of supplies will be key in acceptance of advanced drop-in biofuels into the supply chain once they are demonstrated at a convincing scale. To that end, the Committee is supportive of the collaboration between the Navy, Department of Agriculture and DOE to develop innovative technologies for jet and diesel fuels for military uses. With the Department of Defense as an early adopter of these alternative fuels, the wider marketplace will be more likely to follow.

FY2013 National Defense Authorization Act (H.R. 4310/S. 3254)

H.R. 4310

Two provisions of H.R. 4310, as reported by the House Armed Services Committee (H.Rept. 112- 479 of May 11, 2012), affect the Department of Defense's procurement and production of biofuels:

> SEC. 313. EXEMPTION OF DEPARTMENT OF DEFENSE FROM ALTERNATIVE FUEL PROCUREMENT REQUIREMENT.
>
> Section 526 of the Energy Independence and Security Act of 2007 (P.L. 110-140; 42 U.S.C. 17142) is amended by adding at the end the following: `This section shall not apply to the Department of Defense.'
>
> SEC. 314. LIMITATION ON AVAILABILITY OF FUNDS FOR PROCUREMENT OF ALTERNATIVE FUEL.

a) Limitation- Except as provided in subsection (b), none of the funds authorized to be appropriated by this Act or otherwise made available during fiscal year 2013 for the Department of Defense may be obligated or expended for the production or purchase of any alternative fuel if the cost of producing or purchasing the alternative fuel exceeds the cost of producing or purchasing a traditional fossil fuel that would be used for the same purpose as the alternative fuel.

b) Exception- Notwithstanding subsection (a), the Secretary of Defense may purchase such limited quantities of alternative fuels as are necessary to complete fleet certification for 50/50 blends. In such instances, the Secretary shall purchase such alternative fuel using competitive procedures and ensure the best purchase price for the fuel.

S. 3254

As passed by the Senate on December 4, 2012, S. 3254 does not contain any provisions that would restrict the ability of the Department of Defense to procure or produce biofuels.

Two provisions in S. 3254, as reported by the Senate Armed Services Committee (S.Rept. 112- 173 of June 4, 2012), would have affected the Department of Defense's procurement and production of biofuels. Via floor amendments, both of these provisions were stricken from the version of S. 3254 that was passed by the Senate.

S.Amdt. 2985 was sponsored by Senator Mark Udall and co-sponsored by Senators Murray, Shaheen, Bingaman, Hagan, Kerry, Begich, and Tom Udall. This amendment, which passed 62- 37, struck Section 313 from the version of S. 3254 as reported by the Senate Armed Services Committee. Section 313 would have limited the ability of the Department of Defense to purchase alternative fuels whose costs exceeds that of an equivalent traditional fossil fuel.

S.Amdt. 3095 was sponsored by Senator Hagan, and co-sponsored by Senators Johnson (SD), Murray, Shaheen, Collins, Schumer, Stabenow, Whitehouse, Coons, Udall (NM), Tester, and Udall (CO). This amendment, which passed 54-41, struck Section 2823 from the version of S. 3254 as reported by the Senate Armed Services Committee. Section 2823 would have prohibited the Department of Defense from entering into a contract to plan, design, refurbish, or construct biofuels refinery infrastructure unless specifically authorized by law.

These stricken sections were adopted by the Senate Armed Services Committee as amendments sponsored by Senator Inhofe and Senator McCain, respectively. These amendments were each approved in votes of 13-12. A third amendment, sponsored by Senator Inhofe, failed in the Senate Armed Services Committee on a tie vote of 13-13. This amendment would have exempted the Department of Defense from Section 526 of the Energy Independence and Security Act of 2007 (P.L. 110-140), which prohibits federal agencies from purchasing alternative fuels whose lifecycle greenhouse gas emissions exceed those of conventional fuels. This amendment was similar to Section 313 in H.R. 4310 as reported by the House Armed Services Committee on May 11, 2012, and passed by the House.

These two stricken sections, as they appeared in the version of S. 3254 that was reported by the Senate Armed Services Committee on June 4, 2012, are as follows:

SEC. 313. LIMITATION ON AVAILABILITY OF FUNDS FOR PROCUREMENT OF ALTERNATIVE FUEL.

Limitation- Except as provided in subsection (b), none of the funds authorized to be appropriated by this Act or otherwise made available during fiscal year 2013 for the Department of Defense may be obligated or expended for the production or sole purchase of an alternative fuel if the cost of producing or purchasing the alternative fuel exceeds the cost of producing or purchasing a traditional fossil fuel that would be used for the same purpose as the alternative fuel.

Exception- Notwithstanding subsection (a), the Secretary of Defense may purchase such limited quantities of alternative fuels as are necessary to complete engine or fleet certification for 50/50 blends. In such instances, the Secretary shall purchase such alternative fuel using amounts authorized for research, development, test, and evaluation using competitive procedures and shall ensure the best purchase price for the fuel.

SEC. 2823. PROHIBITION ON BIOFUEL REFINERY CONSTRUCTION.

Notwithstanding any other provision of law, neither the Secretary of Defense nor any other official of the Department of Defense may enter into a contract to plan, design, refurbish, or construct a biofuels refinery or any other facility or infrastructure used to refine biofuels unless such planning, design, refurbishment, or construction is specifically authorized by law.

Regarding Section 313, S.Rept. 112-173 states:

Limitation on availability of funds for procurement of alternative fuel (sec. 313)

The committee recommends a provision that would prohibit the use of funds authorized to be appropriated to the Department of Defense in fiscal year 2013 from being obligated or expended for the production or sole purchase of an alternative fuel if the cost exceeds the cost of traditional fossil fuels used for the same purpose, except for continued testing purposes.

The committee notes that in December 2011, the Defense Logistics Agency, on behalf of the Department of the Navy, purchased 450,000 gallons of biofuels for $12.0 million, which equates to $26.66 a gallon. According to the Department of the Navy it was the single largest purchase of biofuel in government history and was carried out in order to “demonstrate the capability of a Carrier Strike Group and its air wing to burn alternative fuels.” The Department of the Navy noted that, despite the use of operation and maintenance funds for the purchase, the demonstration is deemed a research, development, test, and evaluation (RDTE) initiative as justification for the higher cost per gallon.

The committee also notes that the Vice Chief of Naval Operations testified before the Subcommittee on Readiness and Management Support on May 10, 2012, regarding pressure on readiness accounts from increased fuel prices that “every $1 increase in the price per barrel of fuel results in approximately $31M of additional cost annually above our budgeted level.” Therefore, the high cost of fuel has direct and detrimental impact on other readiness accounts.

The committee strongly supports initiatives undertaken by the Department of Defense to reduce the fuel demand of the operational forces through affordable new technologies that increase fuel efficiency and offer alternative sources of power. But given the pressure placed on current and future defense budgets, the committee is concerned about the use of operation and maintenance funds to pay significantly higher costs for biofuels being used for RDTE efforts. Therefore, the committee directs the Secretary of Defense to develop and promulgate guidance to the military services and defense agencies on the difference between the operational use of alternative fuels versus continued RDTE initiatives. (Pages 80-81)

H.R. 4310/P.L. 113-6

Section 315 of the enrolled FY2013 National Defense Authorization Act (H.R. 4310/P.L. 112- 239) states:

> SEC. 315. LIMITATION ON OBLIGATION OF DEPARTMENT OF DEFENSE FUNDS FROM DEFENSE PRODUCTION ACT OF 1950 FOR BIOFUEL REFINERY CONSTRUCTION.
>
> Amounts made available to the Department of Defense pursuant to the Defense Production Act of 1950 (50 U.S.C. App. 2061 et seq.) for fiscal year 2013 for biofuels production may not be obligated or expended for the construction of a biofuel refinery until the Department of Defense receives matching contributions from the Department of Energy and equivalent contributions from the Department of Agriculture for the same purpose.

Department of Defense, Military Construction and Veterans Affairs, and Full-Year Continuing Appropriations Act, 2013 (H.R. 5856, H.R. 933/P.L. 113-6)

For the 2013 fiscal year, the House Appropriations Committee, in its report (H.Rept. 112-493 of May 25, 2012, on H.R. 5856), states:

> ADVANCED DROP-IN BIOFUEL PRODUCTION
>
> The request [for Defense Production Act purchases] includes $70,000,000 for the construction or retrofit of domestic commercial (or pre-commercial) scale advanced drop-in biofuel plants and refineries. The Committee understands that the Department has allocated $100,000,000 of the $150,000,000 program addition to the fiscal year 2012 Defense Production Act account for this effort and that $70,000,000 of this funding likely will not execute until well into fiscal year 2013 or even into fiscal year 2014. While the Committee is supportive of alternative energy development, in these times of decreasing budgets, it does not seem prudent to stockpile funds so far ahead of need. Accordingly the recommendation provides no funding for this effort in fiscal year 2013. The Committee urges the Secretary of Defense to request this funding in future years when it can execute in a timely manner. (Page 203)

For the FY2013 fiscal year, the Senate Appropriations Committee, in its report (S.Rept. 112-196 of August 2, 2012, on H.R. 5856), provides for an additional $100 million for the DPA fund over the amount requested in the FY2013 budget request. The report states:

> Additional Funding - The Committee recognizes the critical role that the DPA title III program serves in strengthening the U.S. defense industrial base and believes that this work is in the national interest. Therefore, the Committee increases funding for DPA by $100,000,000 over the budget request. The Committee directs that the additional funding be competitively awarded to new initiatives and priority consideration should be given to completion of DPA projects initiated in prior years. Furthermore, the Committee directs the Under Secretary of Defense for Acquisition, Technology, and Logistics to inform the congressional defense committees 30 days prior to any obligation or expenditure of these funds. (Page 164)

The Department of Defense, Military Construction and Veterans Affairs, and Full-Year Continuing Appropriations Act, 2013 (H.R. 933/P.L. 113-6) includes the FY2013 DOD appropriations as Division C. This act appropriates $223,531,000 for the Defense Production Act Fund, an increase of about $134.3 million over the amount requested. However, the explanatory statements of both the House and the Senate, as passed by the Senate on March 20, 2013 and the House on March 21, 2013, reduced the funding requested for production of advanced drop-in biofuels by $10 million as "ahead of need," for an appropriation of $60 million.

H.R. 933/P.L. 113-6 also included $20 million in funding for research into lowering the emissions of coal to liquid fuels, in Research & Development, Air Force. The explanatory statements of the Senate (as passed March 20, 2013) and House (as passed March 20, 2013) state:

> PROMOTING ENERGY SECURITY
>
> The conferees do not include a provision as proposed by the House regarding the Energy Independence and Security Act. However, the conferees provide $20,000,000 in Research,
>
> Development, Test and Evaluation, Air Force only for research that will improve emissions of coal to liquid fuel to enable this technology to be a competitive alternative energy resource to meet the goals established in the Department of Defense's Operational Energy Strategy and its Implementation Plan. The conferees direct the Secretary of the Air Force, in consultation with the Assistant Secretary of Defense for Operational Energy Plans and Programs, to inform the congressional defense committees 30 days prior to any obligation or expenditure of these funds.

Under the Department of Defense, Military Construction and Veterans Affairs, and Full-Year Continuing Appropriations Act, 2013 (H.R. 933/P.L. 113-6), the Department of Energy funding was appropriated under a continuing resolution. DOE's Energy Efficiency and Renewable Energy was funded at approximately $1.8 billion.

End Notes

[1] P.L. 112-81

[2] §902 of the FY2009 Duncan Hunter National Defense Authorization Act established the position of Director of Operational Energy Plans and Programs (OEPP). The FY2011 Ike Skelton National Defense Authorization Act redesignated the position as an Assistant Secretary of Defense. The position is now codified at 10 U.S.C. 138c.

[3] Office of the Assistant Secretary of Defense, Operational Energy Plans and Programs. "Alternative Fuels Policy for Operational Platforms," July 5, 2012. http://energy.defense.gov/2012-7-5_DoD_Alternative_Fuels_Policy_for_Operational_Platforms.pdf. For more information about how lifecycle greenhouse gas emissions of fuels are calculated, please see CRS Report R40460, *Calculation of Lifecycle Greenhouse Gas Emissions for the Renewable Fuel Standard (RFS)*, by Brent D. Yacobucci and Kelsi Bracmort.

[4] As required by §526 of the Energy Independence and Security Act (EISA) of 2007, P.L. 110-140 for purchases specifically of alternative fuels.

[5] §314 of the FY2012 National Defense Authorization Act, P.L. 112-81.

[6] Created by §902 of the FY2009 National Defense Authorization Act, P.L. 110-417. For more details about the operational energy budget certification process, please see CRS Report R42558, *Department of Defense Energy Initiatives: Background and Issues for Congress*, by Moshe Schwartz, Katherine Blakeley, and Ronald O'Rourke.

[7] Required by §331 of the FY2009 National Defense Authorization Act, P.L. 110-417.

[8] Office of the Assistant Secretary of Defense, Operational Energy Plans and Programs. "Alternative Fuels Policy for Operational Platforms," July 5, 2012.

[9] AR 5-5 Study, "Tactical Fuel and Energy Implementation Plan," prepared on behalf of the U.S. Army Sustainment Center for Excellence by Expeditionary Logistics, Inc. and sponsored by the U.S. Army Combined Arms Support Command. September 24, 2010. Available at http://www.dtic.mil/cgi-bin/GetTRDoc?AD=ADA529051

[10] Air Force HQ. "Air Force Energy Plan 2030."

[11] A 50:50 alternative fuel blend is fuel that consists of 50% conventional (i.e., petroleum-based) fuel and 50% alternative (i.e., non-petroleum-based) fuel.

[12] U.S. Air Force Briefing, "Air Force Energy Consumption," March 6, 2012. The original target date for completing testing and certifying had been 2011. U.S. Air Force, "Air Force Energy Plan 2010," December 9, 2009, p. 8.

[13] The Department of the Navy. "The Department of the Navy's Energy Goals." http://www.navy.mil/features/Navy_EnergySecurity.pdf

[14] The Fischer-Tropsch process is a series of chemical reactions can be used to create liquid fuels from coal, natural gas, or biomass feedstocks.

[15] Hydrotreated jet and diesel biofuels are the more common terms for HEFA (**H**ydroprocessed **E**sters, waste oil feedstocks and **F**atty **A**cids) fuels. These fuels can be created from feedstocks that produce natural oils, such as algae, jatropha and camelina, or from waste animal fats.

[16] For discussion of different types of alternative fuels, please see CRS Report R41460, *Cellulosic Ethanol: Feedstocks, Conversion Technologies, Economics, and Policy Options*, by Randy Schnepf; CRS Report R41282, *Agriculture-Based Biofuels: Overview and Emerging Issues*, by Randy Schnepf; or CRS Report R42122, *Algae's Potential as a Transportation Biofuel*, by Kelsi Bracmort.

[17] P.L. 110-140.

[18] Per a letter from ASD(OEPP) Sharon Burke to Sen. Levin, Chairman of the Senate Armed Services Committee, December 2, 2011.

[19] Department of Defense, *Opportunities for DOD use of Alternative and Renewable Fuels: FY10 NDAA Section 334 Congressional Study*, July 18, 2011. p. 4-10.

[20] Department of Defense, *Opportunities for DOD use of Alternative and Renewable Fuels: FY10 NDAA Section 334 Congressional Study*, July 18, 2011. p. 4-10.

[21] Minty Knighton, "A-10 first aircraft to use alcohol-based fuel," *Air Force Print News Today*, July 2, 2012, http://www.af.mil/ news/story_print.asp?id=123308337.

[22] Discussion with Navy energy official, October 2, 2012.

[23] Memorandum of Understanding Between the Department of the Navy and the Department of Energy and the Department of Agriculture, June 2011, p. 2. Available at http://www.rurdev.usda.gov/SupportDocuments/DPASignedMOUEnergyNavyUSDA.pdf

[24] Memorandum of Understanding Between the Department of the Navy and the Department of Energy and the Department of Agriculture, June 2011, p. 2. Available at http://www.rurdev.usda.gov/SupportDocuments/DPASignedMOUEnergyNavyUSDA.pdf

[25] Memorandum of Understanding Between the Department of the Navy and the Department of Energy and the Department of Agriculture, June 2011, pp. 2-3. Available at http://www.rurdev.usda.gov/SupportDocuments/DPASignedMOUEnergyNavyUSDA.pdf

[26] P.L. 100-140.

[27] Enacted as the Consolidated Appropriations Act, 2012, P.L. 112-74.

[28] Office of the Secretary of Defense, "Department of Defense Fiscal Year (FY) 2013 President's Budget Submission, Justification Book Defense: Production Act Purchases," February 2012, p. 2. http://comptroller.defense.gov/defbudget/fy2013/budget_justification/pdfs/02_Procurement/Office_of_Defense_Production_Act_PB_2013.pdf

[29] Department of Energy, "FY2013 Congressional Budget Request.," Vol. 3. Office of the Chief Financial Officer, February 2012. Energy Efficiency and Renewable Energy: Biomass and Biorefinery Systems Program. p. 40.

[30] U.S. Department of Agriculture, U.S. Department of Agriculture, USDA FY2013 Budget Summary and Annual Performance Plan, February 2012. The Commodity Credit Corporation is the funding mechanism for the mandatory farm commodity program payments that farmers receive from the USDA Farm Service Agency, and some of the conservation payments from the Natural Resources Conservation Service and the Farm Service Agency. The Commodity Credit Corporation also is or has been the funding source for a relatively

small subset of USDA programs for foreign trade, bioenergy, rural development, agricultural research, and other programs. The Commodity Credit Corporation has the authority to borrow up to $30b directly from the Treasury or from private lenders, to be repaid, with interest, through appropriations from Congress. This borrowing authority does not require specific congressional appropriations. For more information about the CCC, please see CRS Report R41254, *Defense: FY2011 Authorization and Appropriations*, coordinated by Pat Towell. http://www.obpa.usda.gov/budsum/ FY13budsum.pdf. p. 21.

[31] P.L. 80-806. From the USDA FY2013 Budget Summary and Annual Performance Plan, "Section 4(e) of the CCC Charter Act authorizes CCC to take action to increase the use of agricultural commodities by "...aiding in the development of new and additional markets, marketing facilities, and uses for such commodities." Under this authority, CCC will make available up to $171 million to subsidize the production of bio-based jet fuel. Because there is no existing viable commercial source for the large-scale production of such fuel, CCC has entered into an agreement with the Department of Energy and the Navy to assist in the development of this product." p. 23. For more information regarding USDA's authority to engage in renewable energy programs, please see CRS Report R41985, *Renewable Energy Programs and the Farm Bill: Status and Issues*, by Randy Schnepf.

[32] Department of Defense, FY2013 President's Budget Submission. "Defense Production Act Purchases Justification Book." February 2012. pg. iv. Available at http://comptroller.defense.gov/defbudget /fy2013/budget_justification/pdfs/02_Procurement/ Office_of_Defense_Production_Act_PB_2013.pdf

[33] This office was created as the Director of Operational Energy Plans and Programs via §902 of the FY2009 NDAA (P.L. 110- 417. §902 of the FY2011 NDAA (P.L. 111-383) designated the position as an Assistant Secretary of Defense. This office was given the responsibility of leading and overseeing DOD's alternative fuel activities, issuing guidelines and policy to streamline alternative fuels investment across DOD and making recommendations regarding the development of alternative fuels by the military departments and the Office of the Secretary of Defense by §314 of the FY2012 NDAA (P.L. 112-81).

[34] Department of Energy, "FY2013 Congressional Budget Request.," Vol. 3. Office of the Chief Financial Officer, February 2012. Energy Efficiency and Renewable Energy: Biomass and Biorefinery Systems Program. p. 40.

In: Alternative Fuel Use by the Department of Defense
Editor: Rasmus Klavan
ISBN: 978-1-62808-233-3

Chapter 2

THE NAVY BIOFUEL INITIATIVE UNDER THE DEFENSE PRODUCTION ACT*

Anthony Andrews, Kelsi Bracmort, Jared T. Brown and Daniel H. Else

SUMMARY

The Secretaries of Energy, Agriculture, and the Navy have entered into a Memorandum of Understanding (MOU) to "assist the development and support of a sustainable commercial biofuels industry." The objective of the MOU is the construction or retrofitting of multiple domestic commercial or pre-commercial scale advanced drop-in biofuel plants and refineries. The MOU would support the Navy's goal of deploying a "Green Strike Group" by the end of 2012, and "Great Green Fleet" by 2016 fueled in part with a 50/50 blend of hydrotreated renewable jet fuel (biofuel). The Navy proposes to use authority under the Defense Production Act of 1950 (DPA) to develop a domestic industrial capacity to supply biofuel. In its FY2013 Congressional Budget Request, the Department of Energy (DOE) requested authority to transfer funds to the DPA Fund, offering the justification that it will support the MOU with the technical expertise to move pilot-scale demonstration projects to larger-scale production in support of the Navy's Green Fleet Goal. Agriculture, Energy, and the Navy expect to fund this initiative at $510 million in aggregate over three years.

In the past, Congress has found it in the interest of national defense preparedness for government to assure that a domestic industrial capacity exists to produce fuel. Congress set aside the (now depleted) Naval Oil Reserves and Oil Shale Reserves to provide for the Navy's fuel requirements. Congress later promoted alternative fuel from coal through the U.S. Synthetic Liquid Fuels Act of 1944 to aid the execution of World War II, and to conserve and increase national oil resources. The act authorized the Secretary of the Interior to construct, maintain, and operate plants producing synthetic liquid fuel from coal, oil shale, and agricultural and forestry products. During the Korean War, the DPA authorized the

---

* This is an edited, reformatted and augmented version of the Congressional Research Service Publication, CRS Report for Congress R42568, dated June 22, 2012.

President to have liquid fuels processed and refined for government use or resale, and to make improvements to government- or privately-owned facilities engaged in processing and refining liquid fuels when it would aid the national defense. In 1980, Congress amended the DPA to authorize the President's purchase of synthetic fuels for national defense. Most recently, the Energy Policy Act of 2005 directed the Secretary of Energy, in cooperation with the Secretaries of the Interior and Defense, to develop a program to accelerate the commercial development of strategic unconventional fuels, including but not limited to oil shale and tar sands resources within the United States. Except for exploiting the Naval Oil Reserve, policies that directed alternative fuel development for national defense interests have had to challenge newly discovered petroleum resources that presented clear economic advantages over alternative fuels.

Domestic crude oil production in the United States has increased over the past few years, reversing a decline that began in 1986. The United States is now a net exporter of refined petroleum products. Over the next 10 years, continued development of unconventional oil resources, in combination with the ongoing development of offshore resources in the Gulf of Mexico may push domestic crude oil production to a level not seen since 1994, according to the U.S. Energy Information Administration.

An important policy question for Congress may be whether a domestic biofuel industry is necessary for national defense, and whether proceeding under the authority of the DPA offers the necessary stimulus. A domestic biofuel industry may satisfy concerns for a secure, domestic, alternative fuel source independent of unstable foreign petroleum suppliers. However, adding biofuel to the military's supply chain does not relieve logistical issues with delivering fuel to forward operating areas, where fuel supply issues have been more about vulnerability than availability.

INTRODUCTION

During the spring of 2011, the Secretaries of Energy, Agriculture, and the Navy entered into a Memorandum of Understanding (MOU) to "assist the development and support of a sustainable commercial biofuels industry."[1] The objective is the "construction or retrofit of multiple domestic commercial or pre-commercial scale advanced drop-in biofuel plants and refineries." The refineries would have the capability to produce advanced biofuels as "drop-in" replacements for petroleum-based fuels. Drop-in fuels[2] would be able to substitute for standard petroleum products without alteration to ship or aircraft engines, meet military specifications at a price competitive with petroleum, be readily accessible at geographically diverse locations, and have no significant impact on the supply of agricultural commodities for the production of food. Through this and other undertakings, the Navy intends to demonstrate that it can reduce its dependence on petroleum-based fuels. In accordance with the MOU, the Navy proposes to use some of the authorities of the Defense Production Act (DPA) of 1950 (50 U.S.C. Appx §2061 et seq.), as amended, to develop a domestic industrial capacity and supply of biofuel.

The Navy plans a feasibility demonstration of drop-in biofuels with a "Green Strike Group" fleet exercise by the end of 2012 composed of nuclear vessels and ships powered in part by biofuel, followed by a 2016 "Great Green Fleet" exercise composed of nuclear ships, and surface combatants and aircraft powered in part by biofuel.

The DPA grants various authorities to the President—subsequently delegated to certain Cabinet Secretaries[3]—that they may use to support the domestic production of alternative energy sources. Congressional actions on the budget requests for MOU funding would help define the parameters by which agencies acting under the DPA may support the domestic development of alternative energy sources as well as how the Navy's support of such sources will evolve.

The current debate in Congress regarding DPA is whether its authority should extend to developing an industrial capacity for biofuel.

To better the understanding of the potential benefits and drawbacks to the Navy the following discussion provides a background on the Navy's fleet, fuel requirements, and fuel spending. The Navy, and later DOD, have undertaken a number of past initiatives for securing its fuel supply, some under the authority of DPA. The Navy successfully transitioned from exclusive reliance on petroleum in 1955 when it commissioned the USS *Nautilus*, the world's first nuclear-powered submarine. In past hearings on renewing the DPA, interest in the viability of nuclear-powered Navy and a nuclear power industry were part of the debate.[4]

Background on Naval Fuel Supply and Procurement

The Navy currently lists 395 ships in its inventory—36 nuclear powered warships, 50 conventional warships, and the remaining consisting of various auxiliary and civilian operated Military Sealift Command (MSC) ships (of which 14 are oil tankers—"oilers"). Conventional ships consume diesel fuel marine (DFM or NATO equivalent F-76, also referred to as ship's bunker fuel) which has been used in all shipboard propulsion plants (diesel, gas turbine, and steam boiler) since 1975. Naval aircraft consume primarily JP-5 jet fuel (MIL-DTL-5624 U or NATO code F-44) developed for use in military aircraft stationed aboard aircraft carriers where the risk of fire is a great concern, particularly in the confined spaces of the hanger deck. The fuel is kerosene-based, and has a relatively higher flash point (140 °F) than other aviation turbine fuels (commercial Jet A and or military JP-8). JP-5 is also suitable for use as ship turbine fuel.

To date, the Navy has certified the F-18 Super Hornet, the F-18 legacy Hornet, the MV-22 Osprey, and the MH-60S Seahawk to operate on HRJ-5, a 50/50 blend of hydrotreated renewable fuel (HRJ) and conventional JP-5.[5] Hydrotreated refers to the process of cracking vegetable or animal oil through the addition of hydrogen into kerosene-range jet fuel. The process is similar to refining processes for upgrading petroleum.

Department of Defense (DOD) fuel consumption varies from year to year in response to changes in mission and the tempo of operations. In FY2011, DOD consumed 124.3 million barrels of refined petroleum products domestically, and another 96.6 million barrels overseas.[6] In contrast, the United States consumed slightly more than 6 billion barrels of refined petroleum products during calendar year 2011,[7] of which DOD domestic consumption represented roughly 2%. DOD overseas spending on ships' bunker fuel exceeded domestic spending—$2.7 billion for 7.5 million barrels from 73 ports in 50 countries, versus $239 million for 2.4 million barrels from 67 domestic ports.[8]

The Defense Logistics Agency–Energy (DLA-E) purchases fuel for all of DOD's services and agencies, both in the continental United States (CONUS) and outside (OCONUS). In practice, DLA-E typically awards fuel contracts based on the lowest cost to the point of delivery, typically for lengths of one year in a one-step process under the Defense Working Capital Fund (DWCF). It internally transfers (sells) the fuel to its DOD customers with a small markup to cover its operational costs. DOD customers in turn pay DLA-E through their annual appropriations for Operation and Maintenance (O&M). The DWCF permits DLA-E to take advantage of price breaks for large quantity purchases, and in most years provide DOD customers a stabilized price for all products during that fiscal year. DOD has the capability to meet contingency fuel needs through the petroleum war-reserve stock that DLA-E maintains at its Defense Fuel Supply Points throughout the United States, as well as abroad.[9] DOD relies on both NATO partners and alliances with regional partners to supply fuel for overseas operations.

DLA-E authority to procure fuel extends from authority originally granted to the Navy. Under 10 U.S.C. §7229 (Purchase of Fuel), "... the Secretary of the Navy may, in any manner he considers proper, buy the kind of fuel that is best adapted to the purpose for which it is to be used." Section 7229 superseded 34 U.S.C. §580 which had been interpreted as authorizing the Armed Services Petroleum Purchasing Agency to negotiate contracts for the purchase of fuel, not only when acting as a procuring activity for the Navy, but also when filling the consolidated fuel requirements of the armed forces. However, DLA-E now relies on the general procurement authority under 10 U.S.C. §2304 (Contract: competition requirement), since this gives DOD the authority to buy almost any kind of supply or service.

DOD projects that its budget authority and outlay for fuel will decline 3% from the FY2013 base year and remain essentially level through FY2017.[10] For the period FY2008 through FY2010, the Navy consumed about 10.9 million barrels of JP-5 annually.[11] However, over the same FY2008- FY2010 period, Navy fuel spending declined from $4.75 billion to $3.73 billion.[12] In FY2011, DLA-E spent $1.57 billion to purchase 12.6 million barrels of (Navy) JP-5 jet fuel.[13] In 2011, the U.S. government purchases of F-76 military-grade diesel ranged from $2.19 to $2.75 per gallon, and JP-5 jet propellant ranged from $1.55 to $2.87 per gallon.[14] In 2009, DLA awarded small contracts for hydrotreated renewable fuel (HRJ-5 jet fuel) that ranged in price from $66 to $149 per gallon.[15] In late 2011, the Navy was able to purchase biofuel for roughly $26 per gallon in a bulk procurement of 450,000 gallons.

PREVIOUS DEFENSE-RELATED FUEL PROGRAMS

Congress has previously found it in the interest of national defense preparedness for the federal government to assure some level of industrial capacity to produce both conventional and synthetic fuels. In the early 20th century, Congress set aside the (now depleted) Naval Oil Reserves and Oil Shale Reserves. World War II and Korean War concerns over fuel shortages led to congressionally authorized programs for developing synthetic fuels that lead up to authorizing a Synthetic Fuels Corporation during the 1970s Arab oil embargoes. In more recent legislation, Congress has directed DOD to develop a program to accelerate the commercial development of strategic unconventional fuels (from coal, oil shale or tars sands), examine the military utility of alternative and synthetic fuels, and study how to reduce their life-cycle emissions.

Naval Petroleum and Oil Reserves Set Aside

In the Pickett Act of 1910, Congress initially withdrew potential oil-bearing public lands in California and Wyoming as petroleum sources for the use of the Navy. Executive orders later created three Naval Petroleum and Oil Shale Reserves between 1912 and 1927 by setting aside federal lands believed to contain oil as an emergency reserve. In a 1920 naval appropriation bill, Congress gave the Secretary of the Navy the power "to conserve, develop, use and operate the same in his discretion, directly or by contract, lease, or otherwise, and to use, store, exchange, or sell the oil and gas products thereof, and those from all royalty oil from lands in the naval reserves, for the benefit of the United States."

Petroleum Reserve No. 3 in central Wyoming, the site of the infamous Teapot Dome, embroiled the Navy and Interior Department Secretaries in a 1920s scandal over a noncompetitive lease agreement made with an influential oil company executive. The Teapot Dome scandal ended with the indictment of federal officials involved and, more significantly, transformed federal leasing into a process based on open competitive bids. As the national defense needs changed, Congress opened the oil reserves to private production and eventually sold them. Management of the Naval Petroleum Reserves transferred to the Department of Energy (DOE) and of the Naval Oil Shale Reserves to the Department of the Interior (10 U.S.C. §§7420-7439).

Synthetic Liquid Fuels Development

Congress began promoting alternative fuel from coal through the U.S. Synthetic Liquid Fuels Act of 1944. The act intended to aid the execution of World War II and conserve and increase national oil resources by authorizing the Secretary of the Interior to construct, maintain, and operate plants producing synthetic liquid fuel from coal, oil shale, and agricultural and forestry products.[16] The Bureau of Mines received funding for an 11-year demonstration plant program that ended in 1955. During the Korean War, Section 303 of the Defense Production Act of 1950 (Chap. 932, 64 Stat. 978) authorized the President to have liquid fuels processed and refined for government use or resale, and to make improvements to government- or privately-owned facilities engaged in processing and refining liquid fuels when it would aid the national defense.[17]

During the 1970s, DOE directed a synthetic fuels program toward commercializing coal liquefaction, coal gasification, and oil shale technologies. In 1980, Congress amended Section 305 of the Defense Production Act (P.L. 96-294, Energy Security Act) to authorize the President's purchase of synthetic fuels for national defense. President Carter then directed the Secretary of Defense to determine the quantity and quality of synthetic fuel needed to meet national defense needs.[18] Congress further amended the Defense Production Act to offer financial assistance in producing synthetic fuel from coal, oil shale, tar sands, and heavy oils by establishing the U.S. Synthetic Fuels Corporation (P.L. 96-294, the United States Synthetic Fuels Corporation Act of 1980).[19] The stated goal of the act was reaching a daily synthetic fuels production capacity of 500,000 barrels of oil-equivalent by 1987 and 2 million barrels by 1992.

DOD had become interested in oil shale as an alternative resource for producing quality jet fuel as early as 1951. The Navy and the Naval Petroleum and Oil Shale Reserves Office (NPSRO) started large-scale evaluations of oil shale's suitability for military fuels in the early 1970s, and conducted extensive tests of oil-shale distillates processed into gasoline, JP-4 (a highly volatile military jet fuel reserved for cold weather operations), JP-5 jet fuel, DFM/F-76, and a heavy fuel oil used in military and commercial equipment.[20]

In the late 1970s, the Air Force became interested in evaluating oil shale's suitability for producing JP-4 jet fuel and awarded contracts to develop technology for producing oil shale-derived JP-4 jet fuel.[21] Air Force research may have demonstrated oil shale's unsuitability for making JP-4. At best, oil shale produces mostly middle-range petroleum distillates suitable for upgrading to kerosene and diesel fuel, while JP-4 is closer to lighter, gasoline-naphtha. The Air Force generally phased out JP-4 in the early 1990s in favor of kerosene-based JP-8.[22]

By the mid-1980s, the House began considering a bill to abolish the Synthetic Fuels Corporation with the Synthetic Fuels Fiscal Responsibility Act of 1985 (H.R. 935). In the debate of the bill (H.Rept. 99-196), the Energy and Commerce Committee linked abolishing the Corporation to reducing the federal deficit and viewed purchasing oil for the Strategic Petroleum Reserve as a far more cost effective defense against another oil embargo than subsidizing synthetic fuels. Congressional criticism also focused on conflicts of interest among the Corporation board members, high salaries for staff, lack of interest on the part of private industry, and the possibility of huge subsidies going to profitable oil companies. The minority view noted that as late as 1983, the DOD had certified that it needed synthetic fuel to meet national defense needs. Language rescinding most of the Synthetic Fuels Corporation funding was included in the FY1986 continuing appropriations resolution (H.J.Res. 465, P.L. 99-190).

Strategic Unconventional Fuels Commercial Development

The Energy Policy Act of 2005 (P.L. 109-58, Section 369, Oil Shale, Tar Sand and Other Strategic Unconventional Fuels) directed the Secretary of Energy, in cooperation with the Secretaries of the Interior and Defense, to jointly develop a program to accelerate the commercial development of strategic unconventional fuels, including but not limited to oil shale and tar sands resources within the United States. The act amended Chapter 141 of Title 10, United States Code, by inserting Section 2398a (Procurement of Fuel Derived from Coal, Oil Shale) directing the Secretary of Defense to develop a strategy to use fuel produced, in whole or in part, from coal, oil shale, and tar sands in meeting the fuel requirements of the DOD when the Secretary determines that it is in the national interest. While DOD did not pursue oil shale, it did pursue development of coal-to-liquids technology.[23] This pursuit ended when Congress enacted the Energy Independence and Security Act of 2007 (P.L. 110-140), in which Section 526 (Procurement and Acquisition of Alternative Fuels) effectively proscribed any federal effort to purchase alternative or synthetic fuels with lifecycle greenhouse emissions that exceeded emissions of fuels derived from conventional petroleum; a proscription likely to exclude coal-derived and other unconventional resource-derived fuels from facilities that do not capture and sequester carbon emissions. The provision did permit DOD to procure limited quantities of coal and gas-to-liquids fuels for certification testing.

Alternative Fuels Development

The Duncan Hunter National Defense Authorization Act for Fiscal Year 2009 (P.L. 110-417, Section 334, Study on Alternative and Synthetic Fuels) required the Secretary of Defense to conduct a study on alternative fuels to reduce the life cycle emissions of alternative and synthetic fuels (including coal-to-liquids fuels), and in particular, examine the military utility of domestically produced alternative and synthetic fuels for military operations and for use by expeditionary forces. In response, the Secretary directed the RAND Corporation to study and submit a report to Congress.[24] Generally, RAND found "that a domestic alternative fuel industry could yield large economic profits within the United States. By reducing demand for conventional petroleum, alternative fuel production would also lower world oil prices, countering efforts of certain foreign oil suppliers to control prices by restraining their production."[25] However, RAND further concluded, "There is no direct benefit to the Department of Defense or the services from using alternative fuels rather than petroleum-derived fuels." With regard to hydrotreated renewable oils, RAND found "insufficient evidence that appreciable quantities of hydrotreated renewable oils can be affordably produced in the United States in a manner that does not increase lifecycle greenhouse gas emissions. Oil production from crops, even if they are not food crops, raises important land-use issues. Oil production from meat and poultry processing is inherently limited by the amount of available wastes." RAND considered the prospects for affordably producing fuel from algae in a climate-friendly manner to remain extremely uncertain.

THE DEFENSE PRODUCTION ACT AND ENERGY

The Defense Production Act of 1950 (DPA), as amended,[26] provides the President an "array of authorities to shape *national defense* preparedness programs and to take appropriate steps to maintain and enhance the domestic industrial base" [italics added].[27] As defined, "national defense" establishes the scope of all DPA authorities, and the use of any authority under the DPA must inherently be interpreted to "promote, support, or otherwise be deemed needed or essential for the national defense."[28] "National defense" as defined by statute means:

> programs for military and energy production or construction, military or critical infrastructure assistance to any foreign nation, homeland security, stockpiling, space, and any directly related activity. Such term includes emergency preparedness activities conducted pursuant to title VI of The Robert T. Stafford Disaster Relief and Emergency Assistance Act [42 U.S.C. §5195 et seq.] and critical infrastructure protection and restoration.[29]

Therefore, the purpose of the DPA authorities extends beyond shaping the U.S. military preparedness and capabilities, as the authorities may be used to enhance and support domestic preparedness, response, and recovery from natural hazards, terrorist attacks, and other national emergencies, among other issues. The full purpose of the DPA is discussed in the "Declaration of Policy" issued as part of the statute.[30]

The DPA has been amended and reauthorized numerous times since its original enactment because the statute includes a sunset provision that has required periodic reauthorization and offered the opportunity for amendment. Congress passed the DPA in 1950 and has thus far reauthorized it 17 times, in 1951, 1952, 1953, 1955, 1974, 1975, 1977, 1980, 1984, 1986, 1991, 1992, 1995, 2001, 2003, 2008, and 2009.[31] The authorities provided in the DPA, with few exceptions, are set to expire on September 30, 2014.[32]

How the DPA Applies to Energy

With regard to energy and national defense, the declaration of policy that accompanies the DPA states that "in order to ensure national defense preparedness, it is necessary and appropriate to assure the availability of domestic energy supplies for national defense needs," and "to further assure the adequate maintenance of the domestic industrial base, to the maximum extent possible, domestic energy supplies should be augmented through reliance on renewable energy sources (including solar, geothermal, wind, and biomass sources), more efficient energy storage and distribution technologies, and energy conservation measures."[33]

In Executive Order 13603 on "National Defense Resources Preparedness,"[34] which delegates DPA authorities, the general term "energy" is defined for the purposes of DPA authorities as "all forms of energy including petroleum, gas (both natural and manufactured), electricity, solid fuels (including all forms of coal, coke, coal chemicals, coal liquefaction, and coal gasification), solar, wind, other types of renewable energy, atomic energy, and the production, conservation, use, control, and distribution (including pipelines) of all of these forms of energy."[35]

In other words, under the DPA's declaration of policy, Congress has found that it is in the interest of national defense preparedness that the government assures that some level of capacity exists in the domestic industrial base to produce and provide both traditional and renewable energy sources, including from biomass sources.

Authorities under Title III (Expansion of Productive Capacity and Supply)

Authorities in Title III of the DPA deal with the expansion of national defense productive capacity and supply.[36] The various sections authorize the President to provide or guarantee loans to industry in order to expedite deliveries or expand discovery and production of essential materials, develop technological processes, purchase industrial items or technologies for installation in government or private industrial facilities, make purchases of materials for government use or resale, and provide for the increased use of emerging technologies.

Using Title III authorities, the President may also provide appropriate incentives to develop, maintain, modernize, restore, and expand the productive capacities of domestic sources for critical components, critical technology items, materials and industrial resources essential for the execution of the national security strategy of the United States. Further, the DPA authorizes the President to take appropriate actions to assure that critical components, critical technology items, essential materials, and industrial resources are available from reliable sources when needed to meet defense requirements during peacetime, graduated mobilization, and national emergency.[37]

The DPA statute designates "energy" as one such "strategic and critical" material.[38] However, prior to the DPA being amended and reauthorized in the Defense Production Act Reauthorization of 2009 (Reauthorization of 2009),[39] this designation contained a restriction that the DPA did *not* grant any new direct or indirect authority to the President to "engage in the production of energy in any manner whatsoever (such as oil and gas exploration and development, or any energy facility construction), except as expressly provided in Sections 305 and 306 [50 U.S.C. Appx. §2095 and 2096] for synthetic fuel production." This restriction to the use of DPA authority for energy was removed in Section 5 of the Reauthorization of 2009.[40]

In addition, the Reauthorization Act of 2009 removed the sections of the DPA relating specifically to the production of synthetic fuels (formerly Sections 305, 306, 307, and 308 of the DPA; 50 U.S.C. Appx. §2095 to §2098). Congress no longer considered these provisions necessary, as they served to exempt the synthetic fuels from the restriction on how the DPA could be used for energy production.

Appropriations for Title III of the DPA

Funding for all provisions and purposes of the DPA are authorized in statute.[41] Title III of the DPA also creates a Treasury account named the Defense Production Act Fund (the Fund) that is used to carry out all of the provisions and purposes of Title III. The money in the Fund is available for obligation until expended. The Fund is also used to collect all appropriated funds and proceeds from DPA activities under Title III.[42] However, at the end of any fiscal year, the balance in the Fund cannot exceed $750 million, excluding moneys appropriated that fiscal year or obligated amounts.[43] **Table 1** provides the appropriations to the DPA Fund since FY2010.

Use of Title III Authorities of the DPA for Energy

According to the Defense Production Act Committee, an interagency body created in the Reauthorization of 2009 to coordinate the use of the DPA,[44] the federal government has not used the loan guarantee authority provided in Section 301 of Title III[45] in more than 30 years.[46]

Current projects are initiated under Section 303 of Title III.[47] Section 303 grants the President an array of authorities to create, maintain, protect, expand, or restore domestic industrial base capabilities essential to the national defense. These authorities include, but are not limited to purchasing or making purchase commitments of industrial resources;[48] encouraging the exploration, development, and mining of critical and strategic materials;[49] and installing and purchasing equipment for industrial facilities to expand their productive capacity.[50] Because "energy" is designated as a "critical material," Section 303 of Title III may therefore authorize the President to make provision for the development of different types of energy that are essential for the national defense.

Table 1. Appropriations to the DPA Fund since FY2010, in Millions

Fiscal Year	Law	Appropriation Amount
2010	P.L. 111-118, 123 Stat. 3422	$150.7
2011	P.L. 112-10, 125 Stat. 51	$34.3
2012	P.L. 112-74, 125 Stat. 800	$170.0

Source: CRS analysis of appropriations. Dollars rounded to the nearest hundred thousand.

As with most other authorities the DPA grants to the President, Section 303 of E.O. 13603 delegates authority to the "head of each agency engaged in procurement for national defense."[51] The Secretary of Energy is one such "head of an agency," as defined specifically by E.O. 13603.[52] Though multiple agencies are delegated Title III authorities, DOD was the only agency with an active Title III program as of August of 2011. However, other Departments, including Energy and Homeland Security, have worked with DOD on projects of mutual interest in the past.[53] There are two current Title III projects funded through the DPA Fund and DOD that directly relate to energy.[54] They are a "Lithium Ion Space Battery Production Initiative" and a "Thermal Battery Production" initiative.[55]

Biofuels Memorandum of Understanding

On March 30, 2011, President Obama directed the Secretaries of Agriculture, Energy and the Navy to "investigate how they can work together to speed the development of "drop-in" biofuels substitutes for diesel and jet fuel" as one part of a broader plan to develop and secure the United States' energy resources.[56] Three months after the challenge was issued, the Department of the Navy (Navy), the Department of Energy (DOE), and the Department of Agriculture (USDA) issued a Memorandum of Understanding (MOU) to, in part, "assist the development and support of a sustainable commercial biofuels industry."[57] All three federal agencies have entered into similar MOUs before, with each other and/or with other nations, to spur the development of biofuels.[58]

This MOU is unique in that it focuses solely on commercial-scale development of advanced biofuels and brings in the primary federal agencies involved in the advanced biofuel pathway. An advanced biofuel is essentially any type of biofuel that is non-corn starch ethanol (e.g., algae- based biofuel, biomass-based diesel, cellulosic ethanol).[59] There are three principal federal agencies involved with advanced biofuel research, development, and production.[60] Typically, Congress views USDA as the federal agency that provides resources and support for biofuel feedstock development and feedstock supply (e.g., Biomass Crop Assistance Program). Typically, DOE is viewed as the federal agency that provides resources and support for biofuel conversion technologies (e.g., loan guarantees). For this MOU, the Navy is the customer for the advanced biofuel.

The objective of the MOU is the construction or retrofit of multiple domestic commercial or pre- commercial scale advanced drop-in biofuel plants and refineries. The refineries should have the capability to produce ready drop-in replacement advanced biofuels that meet military specifications at a price competitive with petroleum, be at geographically diverse locations with ready market access, and have no significant impact on the supply of agricultural commodities for the production of food. USDA, DOE, and the Navy are expected to fund this initiative at an aggregate of $510 million over three years, with that aggregate amount divided equally among them (**Table 2**). The MOU notes that the timelines and means of contributing to the initiative are subject to change.

MOU Status Update

Within the first three months of the agreement, USDA, DOE, and the Navy issued a formal announcement of their intent to collaborate with the private sector to help produce drop-in advanced biofuels for military and commercial applications.[61] In a separate parallel action, the Navy and USDA announced soon after the MOU announcement that DLA-E would purchase 450,000 gallons of advanced drop-in biofuel from a blend of non-food waste and algae. According to USDA, this is the single largest purchase of biofuel in government history.[62]

Additionally, there are plans to direct federal dollars to the development of a pilot or commercial- scale facility that will produce military-grade drop-in biofuels.[63] The Navy intends to use the drop-in advanced fuel in a demonstration of a Green Strike Group in 2012 during the Rim of the Pacific Exercise.

USDA committed to $170 million in the MOU, and requested $171 million in the FY2013 President's budget request.[64] The funding is being requested under the CCC Charter Act authority granted to USDA to subsidize the production of bio-based jet fuel.[65] USDA reports that there is no existing viable commercial source for the large-scale production of such fuel, and as such, the CCC has entered into an agreement with DOE and the Navy to assist in the development of this product. The current farm bill debate could influence the amount of support USDA gives to the MOU. The 2008 farm bill (Food, Conservation, and Energy Act of 2008, P.L. 110-246) contains an energy title with roughly a dozen bioenergy programs, all of which expire at the end of FY2012.[66] Because of extreme budget uncertainty, it is difficult to anticipate what the funding levels for these programs within this title will be. Any uncertainty about federal funding could also be a disincentive for the private sector to support the advanced biofuels industry. The current economic climate could preclude significant funding for both energy title programs and USDA funding for the MOU.

FY2013 President's Budget Request

In the FY2013 President's budget request, DOE stated it will support this agreement by taking the technical expertise developed in FY2012 and moving to pilot-scale demonstration projects in FY2013. Up to $40 million may be spent on the pilot-scale demonstration projects to support DOE's ongoing technology maturation program and eventually lead to larger-scale production to support the Navy.[67] The pilot-scale demonstration projects could be chosen from potential joint solicitations for defense biofuels production. In its FY2013 request, DOE seeks conditional legislative language that would allow up to $100 million in new appropriations for the Office of Energy Efficiency and Renewable Energy (EERE) to be transferred to the Defense Production Act Fund for pilot demonstrations of drop-in jet biofuels.

Table 2. MOU Funding Commitment
$ Million

	DPA	CCC	Total
DOE	$170	—	$170
USDA	—	$170	$170
Navy	$170	—	$170
Total	$340	$170	$510

Source: MOU.
Notes:
a. DPA – Defense Production Act; CCC – Commodity Credit Corporation Act.
b. The MOU acknowledges that the funding contributions are subject to the availability of funds and subject to each Department's right to terminate the MOU.
c. Funds are to be transferred in accordance with the Defense Production Act (DPA), (50 U.S.C. App. 2061 et seq.), the Commodity Credit Corporation (CCC) Charter Act (15 U.S.C 714 et seq.), the Economy Act (31 USC 1535); and/or other appropriate authority.

The Navy requested $70 million in the FY2013 President's budget request.[68] The funding would be used to "to execute a DPA Title III project, building upon previous efforts and investments by the Navy and other agencies, with the principal objective being a government-industry partnership for the construction or retrofit of multiple domestic commercial or pre-commercial scale advanced drop-in biofuel plants and refineries." The Navy has collaborated on a regular basis with DOE and USDA.

Defense Production Act Title III Advanced Drop-In BioFuels Production Project

In March 2012, the Air Force issued a Special Notice as defined in the Federal Acquisition regulation (FAR) 5.205 (c) to notify industry that the Defense Production Act (DPA) Title III Program anticipates possible issuance of a Broad Agency Announcement which would request proposals from domestic sources to execute an Advanced Drop-In Biofuel Production Project.[69] The DPA Title III Program would be managed within the Office of the Deputy Assistant Secretary of Defense for Manufacturing and Industrial Base Policy (MIBP) and executed by the Air Force Executive Agent Program Office, a component of the Manufacturing Technology Division (AFRL/RXM) of the Materials and Manufacturing Directorate, Air Force Research Laboratory. As the DOD Executive Agent for the DPA Title III Program, the Air Force is responsible for executing projects that ensure domestic production capability for technology items that are essential to national defense.

The stated goal of the project is establishing one or more complete value chains capable of producing the biofuels. This includes feedstock production, conversion and processing, blending, transportation, and logistics, as well as the design, retrofit, construction, operation, validation, and qualification of domestic, commercial-scale, integrated biorefineries. In the preliminary description provided by the Special Notice, the project would require domestically produced biofuel, compliance with Section 526 of the Energy Independence and Security Act of 2007 (P.L. 110-140), an acceptable feedstock source, suitability for military operational use, a biorefinery with a rated capacity of at least 10 million gallons of neat biofuel per year, and a commitment of at least 50% cost sharing. The project would take a phased approach, with Phase 1 representing completion of planning and preliminary design, with contracts to be tentatively awarded in early 2013. Phase 2 represents facility construction, commissioning, and performance testing as well as delivery of the biofuels, and assumes that awardees would be under contract in early 2014.

FY2013 AUTHORIZATIONS AND APPROPRIATIONS

As of the this report's date, both the Committees on Energy and Water Development, and the Armed Forces have introduced report language regarding the Navy's biofuel initiative. The House Energy and Water Development Committee does not recommend funding, while both Senate Committees for Energy and Water Development and Armed Forces support the initiative.

FY2013 Energy and Water Development Appropriations Bill, H.Rept. 112-462 to Accompany H.R. 5325

The House report to accompany H.R. 5325 offers no funding for the proposed initiative, and it does not include the requested legislative language.[70] Specifically, the Committee notes "the Department has not adequately justified why the Department of Energy should fund this Defense initiative, and whether the proposed investments can successfully lower costs to competitive levels in several years or will only serve to sink costs into a product that is too immature to compete without federal support. The recommendation includes no funding for the proposed initiative and does not include the requested legislative language."

FY2013 Energy and Water Development Appropriations Bill, S.Rept. 112-164 to Accompany S. 2465

The Senate report to accompany S. 2465 supports the collaboration between the Navy, Agriculture, and Energy.[71] Specifically, the Committee "recognizes that quality and reliability of supplies will be key in acceptance of advanced drop-in biofuels into the supply chain once they are demonstrated at a convincing scale.

To that end, the Committee is supportive of the collaboration between the Navy, Department of Agriculture and DOE to develop innovative technologies for jet and diesel fuels for military uses. With the Department of Defense as an early adopter of these alternative fuels, the wider marketplace will be more likely to follow."

FY2013 National Defense Authorization Act, S.Rept. 112-173 Report to Accompany S. 3254

With regard to the limitation on availability of funds for procurement of alternative fuel (Section 313), the Senate Armed Forces Committee "recommends a provision that would prohibit the use of funds authorized to be appropriated to the Department of Defense in fiscal year 2013 from being obligated or expended for the production or sole purchase of an alternative fuel if the cost exceeds the cost of traditional fossil fuels used for the same purpose, except for continued testing purposes.... The committee strongly supports initiatives undertaken by the Department of Defense to reduce the fuel demand of the operational forces through affordable new technologies that increase fuel efficiency and offer alternative sources of power. But given the pressure placed on current and future defense budgets, the committee is concerned about the use of operation and maintenance funds to pay significantly higher costs for biofuels being used for RDTE[72] efforts. Therefore, the committee directs the Secretary of Defense to develop and promulgate guidance to the military services and defense agencies on the difference between the operational use of alternative fuels versus continued RDTE initiatives." The committee also recommends a provision (Section 2823) that would prohibit the Department of Defense from entering into a contract to plan, design, or construct a biofuel refinery or any other facility or infrastructure used to refine biofuels unless such planning, design, or construction is specifically authorized by law.

FY3101 National Defense Authorization Act, H. Rept. 112-479 Report to Accompany H.R. 4310

Section 314—Limitation on Availability of Funds for Procurement of Alternative Fuel would prohibit the use of funds for the production or purchase of any alternative fuel if the cost of producing or purchasing the alternative fuel exceeds the cost of producing or purchasing a traditional fossil fuel. This section would also provide an exception for the Secretary of Defense to purchase limited quantities of alternative fuels to complete fleet certification of 50/50 alternative fuel blends.

POLICY CONSIDERATIONS

Industrial Viability of Advanced Biofuel

Advanced biofuels have been promoted for the past several years as promising potential players in the renewable fuels market. However, there is presently no operating domestic commercial- scale production facility for an advanced biofuel, although some are under construction.[73] Proponents argue that advanced biofuels are the preferred renewable fuel alternative because advanced biofuel feedstocks might not compete with traditional food crops used for food and feed. However, there may still be competition for resources to cultivate the feedstock (e.g., land and water). Proponents of advanced biofuels argue that it will add to the United States' energy security. However, others assert that economic and technical barriers will prevent advanced biofuels from entering mainstream consumption for some time to come. Furthermore, the use of agricultural land for advanced biofuel production was a key sticking point as EPA considered public comment about this aspect of advanced biofuels production eligibility for the Renewable Fuels Standard.[74] Others question whether the federal government should be involved in advanced biofuel production at all.

Some proponents who support advanced biofuels generally question the form that federal involvement should take. For example, some question whether the Navy, using Defense funding, should be the testing ground for advanced biofuels—particularly aviation biofuels—that at present cost significantly more than conventional fuels.[75] Some analysts and government officials counter that argument by reporting that DOD is the single largest consumer of oil in the United States and therefore represents a relatively significant share of total national petroleum use, and that DOD relies "far too much on sometimes politically unstable foreign sources of fossil fuels."[76]

The advanced biofuel industry appears to be most focused on demonstrating that the technology exists to produce advanced biofuels at commercial scale. Another important consideration for the advanced biofuel industry at this moment is the planning and development of other production factors (e.g., feedstock supply, infrastructure) which will play a role in a stable advanced biofuel market—the ultimate goal for the industry.

Biofuel as a National Defense Need

Generally, military/operational fuel supply problems present logistical challenges that may fall under the process of peacetime planning or "deliberate planning"—the process of preparing contingency plans for military action.[77] A domestic biofuel industry, fostered through DPA authorities, would appear to address the contingency of a long-term petroleum supply disruption, assuming that North American production and resources have declined precipitously and shifted reliance to imported crude oil. At present, the United States imports 49% of its petroleum, which includes crude oil (roughly 40%) and refined products.[78] Canada provides the United States with 25% of its imports, and could likely increase the amount depending upon the approval of the Keystone XL pipeline.[79]

However, in the case of short-term disruptions, the United States maintains the Strategic Petroleum Reserve as part of an obligation to the International Energy Agency (IEA) agreement.[80] The SPR's utility in meeting Department of Defense (DOD) needs is limited by volume of jet or diesel fuel a refinery can produce from a barrel of oil. On average, Gulf Coast refineries yield only 6% to 8% kerosene-jet fuel per barrel of oil.[81] That is, a 42-gallon barrel of oil yields at most 3 to 4 gallons of jet fuel when refined. Were it not for motorists, truckers, and airlines purchasing most of the remaining products refined from a barrel of crude oil, the cost of refining jet fuel exclusively for military use would be considerably higher. (This would assume little or no value in upgrading the leftover portion of the barrel). That is, consumers, not taxpayers, effectively subsidize the military jet fuel production.

The policy question writ large is whether a biofuel industry is necessary for national defense. If the answer is yes, then the Defense Production Act may be the appropriate authority for the Navy to proceed with its MOU. Proponents of a domestic biofuel industry view it as necessary in guaranteeing a fuel supply for the needs of the Navy, presuming that the future availability of both domestic and international petroleum resources are may be uncertain. Projections of North American fossil energy supply would call the proponent's argument into question. Just a few years ago, fears of "Peak Oil" production drove policy discussions on U.S. energy security. However, domestic crude oil production has increased over the past few years, reversing a decline that began in 1986. The United States is now a net exporter of refined petroleum products. The U.S. Energy Information Administration (EIA) reports that U.S. crude oil production increased from 5.1 million barrels per day in 2007 to 5.5 million barrels per day in 2010.[82] Over the next 10 years, EIA sees continued development of tight oil, in combination with the ongoing development of offshore resources in the Gulf of Mexico, pushing domestic crude oil production in the EIA Reference Case to 6.7 million barrels per day by 2020, a level not seen since 1994. Even with a projected decline after 2020, EIA projects that U.S. crude oil production will remain above 6.1 million barrels per day through 2035. U.S. refineries consumed just over 5.4 billion barrels of crude oil in 2011, or approximately 14.8 million barrels per day.[83]

Biofuel proponents would also argue that spending U.S. dollars on imported crude oil supports foreign adversaries or props up corrupt regimes. However, out of all the major oil exporting nations, currently the U.S. State Department lists only Iran as a terrorist country, and the United Nations Security Council trade sanctions in place (UNSC Resolution 1929) significantly restricts Iran's oil exports.[84]

No matter whether fuels are domestically or foreign sourced, certain logistical challenges arise with delivering or acquiring fuel in overseas operational areas. For example, during U.S. operations in Afghanistan, insurgent forces frequently attacked over-the-road fuel convoys and that caused numerous U.S. and contractor casualties as well as equipment and fuel losses. Despite the fuel convoys' vulnerability, DLA-E generally found fuel sources to be available; though the sources' alliances and loyalties may have been questionable, the transactions may have gained some regional cooperation. Refueling in foreign ports also introduces certain vulnerabilities, as tragically illustrated by the al-Qaeda attack on USS *Cole* at Aden, Yemen, in October of 2000; the unforeseen consequence of State Department efforts to cultivate U.S.-Yemeni ties. Relying solely on at-sea refueling for fleet units worldwide would also introduce challenges requiring a substantial increase in the number of Military Sealift Command oilers in service. For routine operations, the Navy can expect to keep only about one third of its 14-oiler force at sea at any given moment. The four or five oilers routinely deployed cannot indefinitely sustain a seagoing fleet of three to four carrier battle groups and three to four amphibious readiness groups deployed worldwide, plus a number of naval units steaming independently. Therefore, decisions on whether to refuel Navy ships at sea or in foreign ports depends on both the tactical calculus of committing scarce fleet oilers and the political calculus of "showing the flag" by sending fleet units into port. To some extent, then, this offers an opportunity for the exercise of relatively inexpensive diplomatic "soft power" by the United States in the form of contracts with local companies for the supply of marine fuel.

DOD might argue that it faces financial, operational, as well as strategic risks from its reliance on petroleum-based fuels. The financial risks include rising fuel costs over the long term, and increasing price volatility in the short term. Operational risks include the diversion of resources to the task of moving fuel to the battlefield, fuel requirements restraining the mobility U.S. forces and the combat effectiveness of U.S. equipment, and the vulnerability of fuel supply lines to disruption.

Strategic challenges include getting fuel to the area of operations, and DOD's role in ensuring the global free flow of oil. The Navy might argue that investing in a domestic biofuels industry would insulate the Navy from the effects of volatile crude oil prices, decrease its reliance on foreign oil exports from politically unstable or hostile regimes, as well as enhance U.S. energy independence and energy security by creating a reliable energy source. In testimony before the Senate Subcommittee on Readiness and Management Support, the Vice Chief of Naval Operations underscored the pressure on readiness accounts from increased fuel prices, adding that "every $1 increase in the price per barrel of fuel results in approximately $31M of additional cost annually above our budgeted level."[85]

The Navy has a lower dependence on petroleum than other services because its aircraft carriers and submarines are nuclear-powered. In FY2010, the Navy met 59% of its overall energy needs from petroleum, 22% from nuclear-powered ships, and 19% from electricity (for shore operations). The Navy nuclear reactor program, an outcome of the Defense Production Act, not only validated nuclear power, but also paved the way for commercial nuclear power and provided the first trained reactor operators. With the exception of the Navy's nuclear power program, policies designed to replace petroleum-based fuels with fuels derived from alternative resources in the past have given way when newly discovered petroleum resources presented clear economic advantages. That risk remains even if the Biofuel MOU is funded.

Whatever advantages biofuels may convey, the Navy is not without recourse in acquiring them directly by the customary acquisition process, as limited quantity procurements have already demonstrated. At $26 per gallon, the $510 million the Navy seeks to fund the MOU would purchase 467 thousand barrels of HRJ-5. Assuming that private interests successfully developed biofuel plants under DPA funding, they would still have to submit competitive bids on future fuel contracts under current procurement rules.

End Notes

[1] *Memorandum of Understanding Between the Department of the Navy and the Department of Energy and the Department of Agriculture*, March 30, 2011, http://www.rurdev.usda.gov/SupportDocuments/DPASigned MOUEnergyNavyUSDA.pdf. For more information on bio-based diesel, see CRS Report R41631, *The Market for Biomass-Based Diesel Fuel in the Renewable Fuel Standard (RFS)*, by Brent D. Yacobucci, *The Market for Biomass-Based Diesel Fuel in the Renewable Fuel Standard (RFS)*, by Brent D. Yacobucci. For more information on algae-based biofuel, a commonly mentioned source of potential bio-jet fuel, see CRS Report R42122, *Algae's Potential as a Transportation Biofuel*, by Kelsi Bracmort.

[2] Drop-in fuels are a type of advanced biofuels that may be blended with conventional fuel and used in existing vehicles, and existing transportation and distribution infrastructure.

[3] See Exec. Order No. 13603, 77 Fed. Reg. 16651 (March 22, 2012).

[4] U.S. Congress, House Banking and Currency, *To Renew the Defense Production Act of 1950, As Amended*, 15683, 90th Cong., p. 62.

[5] Chris Tindal (Director for Operational Energy) *Department of Navy Biofuel Initiatives*, October 5, 2011, South Dakota State University Industry Forum, http://ncsungrant.sdstate.org/upload/SDSU_Industry_Forum_CTindal_5Oct2011_v3.pdf.

[6] Defense Logistics Agency-Energy, *FY2011 Fact Book*: "Bulk Product Transport within the Continental United States" (p. 49), "Domestic Into-plane Contracts (p. 36), Domestic Ships' Bunkers Contracts" (p. 37), "Domestic Posts, Camps and Stations Contracts" (p.38), http://www.energy.dla.mil.

[7] U.S. Energy Information Administration, *Petroleum & Other Liquids*, Product Supplied, Area: U.S, Period-Unit: Annual-Thousand Barrels, 2011, http://www.eia.gov/dnav/pet/ pet_cons_psup_dc_nus_mbbl_a.htm.

[8] The DLA-E Ship's Bunkers program provides commercial ship propulsion fuels for military and other U.S. government ships at commercial and military ports worldwide. DLA-E *FY2011 Fact Book*, p. 37.

[9] See CRS Report R40459, *Department of Defense Fuel Spending, Supply, Acquisition, and Policy*, by Anthony Andrews.

[10] *National Defense Budget Estimates for FY2013*, Office of the Undersecretary of Defense (Comptroller), Table 5-5 "Department of Defense Deflators–TOA", pp. 59-60, March 2012, http://comptroller.defense.gov/ defbudget/fy2013/ FY13_Green_Book.pdf.

National Defense Budget Estimates for Fuel (FY2013 Base Year)

	FY11	FY12	FY13	FY14	FY15	FY16	FY17
Fuel Purchase Inflation Rate Assumptions – Outlays	24.2	15.3	-3.1	-12.9	-0.3	0.3	0.6
Fuel Purchase Inflation Rate Assumptions – Budget Authority	24.2	15.3	-3.1	-12.9	-0.3	0.3	0.6
Fuel Deflator - Total Obligational Authority	89.5	103.2	100.0	87.1	86.8	87.1	87.6
Fuel Deflator – Budget Authority	89.5	103.2	100.0	87.1	86.8	87.1	87.6
Fuel Deflator – Outlays	89.5	103.2	100.0	87.1	86.8	87.1	87.6

Source: Green Book Tables 5-2 though 5-9.

[11] Defense Logistics Agency—Energy, *Fact Book Fiscal Year 2010*, p. 24, http://www.desc.dla.mil/DCM/ DCMPage.asp?PageID=721.

[12] *Fact Book FY2010*, p. 22.

[13] Defense Logistics Agency—Energy, *FY2011 Fact Book*, pp. 28-29, http://www.energy.dla.mil.

[14] DLA-E Bulk Fuels Summary of Awards, *Solicitation SP0600-11-R-708, and SP0600-R-0061*, http://www.desc. dla.mil/DCM/DCMPage.asp?pageid=241.

[15] DLA-E Bulk Fuels Summary of Awards, *Solicitation SP0600-09-070,* http://www.desc.dla.mil/DCM/ DCMPage.asp? *pageid=241.*

[16] 30 U.S.C. Sections 321 to 325 authorized $30 million over five years for the construction and operation of demonstration plants to produce synthetic liquid fuels from coal, oil shales, agricultural and forestry products, and other substances.

[17] Ch. 932, 64 Stat. 798 (Title III Expansion of Production Capacity and Supply) intended to develop and maintain whatever military and economic strength necessary to support collective action through the United Nations. The act authorized the diversion of certain materials and facilities from civilian to military use when expansion of production facilities beyond the levels needed to meet civilian demand was required.

[18] The American Presidency Project, Executive Order 12242 Synthetic Fuels, http://www.presidency.ucsb.edu/ ws/index.php?pid=45171, signed September 30, 1980, was later revoked by President Reagan's Executive Order 12346, February 8, 1982. Available through the National Archives at http://www.archives.gov/federal-register/executive- orders/disposition.html.

[19] Title I, Part B of the Energy Security Act of 1980.

[20] Tosco (an acronym for "The Oil Shale Company") was contracted to produce and process 10,000 barrels of oil-shale distillates. Development Engineering, Inc., leased the federal Anvil Points site (Naval Oil Shale Reserve 3) in 1972 and formed the Paraho Development Corporation in 1973 (a consortium of 17 energy companies). Paraho's plans included a five-year program to develop two pilot scale retorts and produce oil-shale distillates for the Navy fuel testing. Paraho initially produced 10,000 barrels of oil-shale distillates that Sohio processed into gasoline, JP-4 and JP-5 jet fuel, DFM/F-76, and a heavy fuel oil at the Gary Western Refinery in Fruita, CO. Though the fuels produced were off- specification, analysis indicated that the refining process could be optimized to produce specification fuels. Paraho was awarded a follow-on contract to produce 100,000 barrels of oil-shale distillates for processing specification fuels in Sohio's Toledo Refinery.

[21] Under Project Rivet Shale, in 1979, the Air Force awarded contracts to Ashland Research and Development, Suntech, Inc., and UOP, Inc., to develop technology to produce oil shale-derived JP-4 jet fuel. In 1982, over 10,000 gallons of JP-4 were processed at the Caribou Four Corners Refinery in Woods Cross, UT, from crude oil-shale distillates produced by Geokinetics. JP-4 specification fuel was produced from other oil shale retorting techniques pioneered by Occidental, Paraho, and Union Oil. Unocal (formerly Union Oil Company) operated the Parachute Creek oil shale plant and reportedly produced 4.6 million barrels of oil-shale distillates from 1985 to 1990 for Air Force evaluation under Project Rivet Shale.

[22] *For further information on oil shale, refer to CRS Report RL33359, Oil Shale: History, Incentives, and Policy*, by Anthony Andrews.

[23] CRS Report RL34133, *Fischer-Tropsch Fuels from Coal, Natural Gas, and Biomass: Background and Policy*, by Anthony Andrews and Jeffrey Logan.

[24] James T. Bartis and Lawrence Van Bibber, *Alternative Fuels for Military Applications*, RAND National Defense Research Institute, ISBN 978-0-8330-5000-7, 2011.

[25] RAND, P. 81.

[26] 50 U.S.C. Appx §2061 et seq.

[27] 50 U.S.C. Appx. §2062(a)(4); Section 2(a)(4) of the DPA (emphasis added).

[28] There are various iterations of this tie to national defense throughout the DPA. Some examples: Title I, Section 101 priorities and allocations authority requires the President to deem action as "necessary or appropriate to promote the national defense" (50 U.S.C. Appx. §2071(a)); Title III authorities can be used when "essential for the national defense" (50 U.S.C. Appx. §§2091(a), 2092(a), 2093(a)); and Title VII Voluntary Agreement authority requires that the use helps "provide for the national defense" (50 U.S.C. Appx. §2158(c)(1)).

[29] 50 U.S.C. Appx. §2152(14); Section 702(14) of the DPA. Further reference can be made to Title VI of the Stafford Act, at 42 U.S.C. §5195(a)(3) for a definition of "emergency preparedness" activities. It states that emergency preparedness "... means all those activities and measures designed or undertaken to prepare for or minimize the effects of a hazard upon the civilian population, to deal with the immediate emergency conditions which would be created by the hazard, and to effectuate emergency repairs to, or the emergency restoration of, vital utilities and facilities destroyed or damaged by the hazard."

[30] 50 U.S.C. Appx. §2062; Section 2 of the DPA. This section includes congressional findings, Section 2(a), and a statement of policy of the United States, Section 2(b).

[31] Most notably, Congress allowed Titles II, IV, V, and VI of the DPA to expire in 1953 with the passage and enactment of P.L. 85-95.

[32] See 123 Stat. 2006, 50 U.S.C. Appx. §2166.

[33] See 50 U.S.C. Appx. §2062(a)(5) and §2062(a)(6).

[34] Exec. Order No. 13603, 77 Fed. Reg. 16651 (March 22, 2012).

[35] Section 801(b) of E.O. 13603.

[36] 50 U.S.C. Appx. §§2091-2094; Title III of the DPA.

[37] 50 U.S.C. Appx. §2077; Section 107(a) of the DPA.

[38] 50 U.S.C. Appx. §2076; Section 106 of the DPA.

[39] P.L. 111-67.

[40] 123 Stat. 2009.

[41] 50 U.S.C. Appx. §2161; Section 711 of the DPA.

[42] See Section 304 of the DPA, 50 U.S.C. Appx. §2094. Examples of proceeds generated from Title III activities could include interest payments on government loans to private business enterprises or the resale of goods purchased under Title III.

[43] Section 304(e) of the DPA, 50 U.S.C. Appx. §2094(e).

[44] See 123 Stat. 2019 to 2020 for the creation of the DPAC in statute. The DPAC is now authorized in Section 722 of the DPA, 50 U.S.C. Appx. §2171.

[45] 50 U.S.C. Appx. §2091.

[46] See the sub-heading "Credit Assistance" at http://www.dpacommittee.com/dpa.htm#titleIII.

[47] 50 U.S.C. Appx. §2093.

[48] Section 303(a)(1)(A) of the DPA; 50 U.S.C. Appx. §2093.

[49] Section 303(a)(1)(B) of the DPA.

[50] Section 303(e).

[51] See Section 303 of E.O. 13603.

[52] See the definition of "head of each agency engage in procurement for national defense" in Section 801(h) of E.O. 13603. The Secretary of Energy is identified in Section 201 of the E.O., and therefore is one of the defined heads of an agency.

[53] Department of Homeland Security, *The Defense Production Act Committee: Report to Congress*, Washington, D.C., August 2011, p. 10.

[54] For a current list of all DPA Title III projects, see Reauthorization of 2009.

[55] See http://www.dpatitle3.com/dpa_db/project.php?id=67 and http://www.dpatitle3.com/dpa_db/ project.php?id=83, respectively.

[56] U.S. President (Obama), "Blueprint for a Secure Energy Future," March 30, 2011, p. 23.

[57] *Memorandum of Understanding Between the Department of the Navy, the Department of Energy, and the Department of Agriculture*, June 2011, http://www.rurdev.usda.gov/SupportDocuments/DPASigned MOUEnergyNavyUSDA.pdf.

[58] There have been at least three biofuel MOUs over the last five years involving the USDA, DOE, or the Navy:

- *Memorandum of Understanding Between U.S. Department of Energy and U.S. Department of Defense*, July 22, 2010, http://energy.gov/sites/prod/files/edg/media/ Enhance-Energy-Security-MOU.pdf.
- U.S. Department of Agriculture, "USDA, Navy Sign Agreement to Encourage the Development and Use of Renewable Energy," press release, January 21, 2010, http://www.usda.gov/wps/portal/usda/usdahome?contentidonly=true&contentid=2010/01/0029.xml.
- U.S. Department of Agriculture, "U.S. and China Increase Biofuels Cooperation Ahead of the Third U.S. - China Strategic Economic Dialogue," press release, December 12, 2007, http://www.usda.gov/wps/portal/usda/usdahome?contentidonly= true&contentid=2007/12/0370.xml.

[59] Under the Renewable Fuel Standard (RFS), as amended by the Energy Independence and Security Act Policy Act of 2007, gasoline and diesel fuel must contain a minimum amount of fuel produced from renewable biomass, advanced biofuels are biofuels produced from non-corn feedstocks that must reduce lifecycle greenhouse gas

emissions by 50% relative to conventional fuel to qualify. For more information, see CRS Reports CRS Report R40155, *Renewable Fuel Standard (RFS): Overview and Issues*, by Randy Schnepf and Brent D. Yacobucci. For more information on algae- based biofuels, see CRS Report R42122, *Algae's Potential as a Transportation Biofuel*, by Kelsi Bracmort. For more information on cellulosic biofuel, see CRS Report R41106, *Meeting the Renewable Fuel Standard (RFS) Mandate for Cellulosic Biofuels: Questions and Answers*, by Kelsi Bracmort.

[60] Other federal agencies that sit on the periphery of advanced biofuel production are the Environmental Protection Agency (EPA) and the Energy Information Administration (EIA). These two agencies are more concerned with the volume produced, if any, for RFS compliance.

[61] U.S. Department of Agriculture, "USDA, Departments of Energy and Navy Seek Input from Industry to Advance Biofuels for Military and Commercial Transportation," press release, August 30, 2011, http://www.usda.gov/wps/portal/usda/usdahome?contentid=2011/08/ 0385.xml&contentidonly=true.

[62] U.S. Department of Agriculture, "Navy Secretary Ray Mabus and USDA Secretary Tom Vilsack Announce Largest Ever Government Purchase of Biofuel," press release, December 5, 2011, http://usda.gov/wps/portal/usda/usdamediafb?contentid=2011/12/0500.xml& printable=true&contentidonly=true.

[63] The White House, *The Blueprint for A Secure Energy Future: Progress Report*, March 2012, http://www.whitehouse.gov/sites/default/files/email-files/the_blueprint_for_a_secure_energy_future_oneyear_progress_report.pdf.

[64] U.S. Department of Agriculture, *USDA FY2013 Budget Summary and Annual Performance Plan*, February 2012, http://www.obpa.usda.gov/budsum/FY13budsum.pdf.

[65] The CCC provides funding for commodity programs administered by the Farm Service Agency and many Farm Bill programs such as the conservation programs administered by FSA and the Natural Resources Conservation Service (NRCS) and export programs administered by the Foreign Agricultural Service (FAS). CCC borrows funds needed to finance these programs from the U.S. Treasury and repays the borrowings, with interest, from receipts and from appropriations provided by Congress.

[66] For more information, see CRS Report R41985, *Renewable Energy Programs and the Farm Bill: Status and Issues*, by Randy Schnepf.

[67] U.S. Department of Energy, *FY2013 Congressional Budget Request Volume 3*, DOE/CF-0073, February 2012, http://www.cfo.doe.gov/budget/13budget/Content/Volume3.pdf.

[68] U.S. Department of Defense, *FY2013 President's Budget Submission*, February 2012, http://comptroller.defense.gov/defbudget/fy2013/budget_justification/pdfs/02_Procurement/Office_of_Defense_Production_Act_PB_2013.pdf.

[69] U.S. Air Force, *Defense Production Act Title III Advanced Drop-In Bio Fuels Production Project*, May 15, 2012, https://www.fbo.gov/index?s=opportunity&mode=form&tab=core&id=9ec4273b98ce62c3178916a432967581&_cview=0.

[70] H.Rept. 112-462, page 86.

[71] S.Rept. 112-164, page 73.

[72] Research, Development, Testing, and Evaluation.

[73] For instance, POET LLC's Project Liberty plant (http://www.projectliberty.com/).

[74] For more information, see CRS Report R40460, *Calculation of Lifecycle Greenhouse Gas Emissions for the Renewable Fuel Standard (RFS)*, by Brent D. Yacobucci and Kelsi Bracmort.

[75] Fred Lucas, "Navy Biofuel Deal is 'Cost Prohibitive,' 'Another Solyndra,' Critics Say," CNSNews.com, December 23, 2011.

[76] U.S. Navy, Department of Navy Biofuel Initiatives, Biofuels and the U.S. Military webinar, April 4, 2012, http://www.acore.org/wp-content/uploads/2012/03/Chris-Tindal-Presentation.pdf.

[77] Contingency planning for military action begins with the President's national strategy statement backed with the necessary funding support provided by Congress. See Joint Forces Staff College, *The Joint Staff Officer's Guide 2000*, JFSC Pub 1, 2000, Chapter 4-Deliberate Planning, http://www.au.af.mil/au/awc/awcgate/pub1/index2000.htm.

[78] U.S. Energy Information Administration, *How Dependent Are We on Foreign Oil?*, http://www.eia.gov/energy_in_brief/foreign_oil_dependence.cfm.

[79] CRS Report R41668, *Keystone XL Pipeline Project: Key Issues*, by Paul W. Parfomak et al., and CRS Report R42124, *Proposed Keystone XL Pipeline: Legal Issues*, by Adam Vann et al.

[80] See CRS Report R42460, *The Strategic Petroleum Reserve: Authorization, Operation, and Drawdown Policy*, by Anthony Andrews and Robert Pirog.

[81] See CRS Report R41478, *The U.S. Oil Refining Industry: Background in Changing Markets and Fuel Policies*, by Anthony Andrews, Robert Pirog, and Molly F. Sherlock, *The U.S. Oil Refining Industry: Background in Changing Markets and Fuel Policies*, by Anthony Andrews, Robert Pirog, and Molly F. Sherlock.

[82] U.S. Energy Information Administration, *Annual Energy Outlook 2012 Early Release Overview*, January 23, 2012, p. Executive Summary, http://www.eia.gov/forecasts/aeo/er/ executive_summary.cfm.

[83] U.S. Energy Information Administration, "U.S. Refinery and Blender Net Input of Crude Oil" in *U.S. Crude Oil Supply and Disposition*, http://www.eia.gov/dnav/pet/pet_sum_ crdsnd_k_a.htm.

[84] U.S. Department of State, Directorate of Defense Trade Controls, *Country Policies and Embargoes*, http://pmddtc.state.gov/embargoed_countries/index.html.

[85] U.S. Congress, Senate Committee on Armed Services, *National Defense Authorization Act for Fiscal Year 2013*, Report to Accompany S. 3254, 112th Cong., 2nd sess., June 4, 2012, 112-173, pp. 80-81.

In: Alternative Fuel Use by the Department of Defense
Editor: Rasmus Klavan ISBN: 978-1-62808-233-3

Chapter 3

OPPORTUNITIES FOR DOD USE OF ALTERNATIVE AND RENEWABLE FUELS: FY10 NDAA SECTION 334 CONGRESSIONAL STUDY*

U.S. Department of Defense

EXECUTIVE SUMMARY

Section 334 of the National Defense Authorization Act (NDAA) for Fiscal Year 2010 requires the Department of Defense (DoD) to provide an assessment of the use of renewable fuels in non-tactical and tactical aviation, maritime, and ground transportation fleets and asks whether establishing a DoD commodity class for renewable fuels distinct from petroleum-based products would be beneficial. The Office of the Assistant Secretary of Defense for Operational Energy Plans and Programs and the Defense Logistics Agency, in conjunction with the military Services, prepared this report.

This report was prepared in response to the Section 334 requirement and assesses renewable fuel supply (anticipated feedstock availability, production capacity and production) and demand (projected fuel quantities based on the military Services' requirements and plans) through 2020. To do so, the report reviews statutory, regulatory, and other drivers relevant to DoD use of alternative and renewable fuels; discusses relevant DoD and Service policies, programs, and goals; projects the rapidly changing U.S. domestic renewable fuels market; and examines the implications for DoD renewable fuel use by 2020.

The major findings of the report are as follows:

- Increased DoD renewable fuel use helps advance U.S. strategic energy security interests, achieve the Services' goals, and gain some limited military utility, such as lower freeze points, cleaner combustion, and potential for designer fuels.
- At present, these fuels command a price premium, but it is anticipated to decline significantly as the market develops over the next decade. Despite this reduced

---

* This is an edited, reformatted and augmented version of the U.S. Department of Defense, dated July 18, 2011.

premium, the Services' renewable fuel goals could still impose $2.2 billion in additional estimated annual fuel costs by 2020.

- There also are questions of renewable drop-in fuel availability relative to demand. DoD would require more than 40 percent of the total projected U.S. drop-in renewable fuel supply (regardless of fuel type) in 2020, just to meet the military Services' stated demand for 745 million gallons.
- Drop-in renewable jet fuel production is not likely to meet the Services' goal-based demand for more than 570 million gallons in 2020.
- EISA's RFS2 excludes jet fuel, DoD's primary operational fuel, from volumetric production mandates, reducing private-sector incentive to produce renewable alternatives to conventional jet fuel.
- Camelina appears to be a promising renewable feedstock for producing hydrotreated renewable jet (HRJ) fuel, but annual production capacity for camelina-based HRJ is projected at only 68–98 million gallons by 2020.
- Third generation renewable fuels production systems, such as photosynthetic algae, are unlikely to supply significant quantities of feedstock oil by 2020 and may involve consideration of the water requirement tradeoffs.
- The additional costs and potential adverse effects of creating a new DoD commodity class outweigh the potential benefits.

Over the last decade, DoD's use of renewable fuel in non-tactical vehicles (NTVs) has been driven by the Services' sustainability programs and compliance with federal fleet alternative fuel statutory and executive order requirements. Although DoD's renewable fuel consumption continues to grow, it has been limited to supplanting the use of petroleum in its NTVs, primarily with ethanol and biodiesel.

DoD's largest opportunity for renewable fuel use is in its tactical systems and weapons platforms, which constitutes 90 percent of its petroleum fuel demand. These systems are replaced on a generational scale, so renewable fuels used in these systems must conform to existing fuel specifications and performance requirements. Prior to use in tactical systems, renewable fuels must be qualified, and weapon platforms certified, to ensure the fuel does not compromise mission performance or safety. Also, because of DoD efforts to simplify fuel logistics through the use of a single battlefield fuel (JP-8), proposed renewable fuels that require separate supply chains or are incompatible with existing infrastructure may face military doctrinal challenges.

However, renewable fuels also offer the means to reduce U.S. force dependence on petroleum fuels. For this reason, DoD has been exploring their use in its tactical applications through the Services' comprehensive test and evaluation programs and plans to procure substantial volumes over time (745 million gallons annually by 2020, or 14 percent of operational fuel). The Services' technical communities are aggressively testing and qualifying drop-in renewable fuels. They have developed qualification or certification road maps and are working on the ability to use these drop-in renewable fuels in large numbers of tactical systems and weapons platforms by 2016. So far, they have determined that renewable fuel blends (with conventional petroleum) that meet military diesel and jet fuel specifications will not require separate infrastructure or pose maintenance risks to existing assets.

DoD faces two major challenges in meeting the Services' goals for renewable fuel use:

1. *Ensuring a sufficient supply of drop-in renewable fuel, particularly jet fuel.* The aggregate supply of drop-in renewable (jet and diesel) fuel may not meet both DoD and commercial demand. Given the Services' goals and projected supply, DoD would have to capture more than 40 percent of the renewable and cellulosic diesel and jet markets in 2020. The Services' 2020 goals for renewable jet fuel alone far exceed even the high-end projected domestic supply (Figure ES-1).
2. *Providing drop-in renewable fuel at an acceptable cost.* Drop-in renewable fuels are expected to cost more than their petroleum counterparts: the estimated price premium will be between $1.43 and $5.24 per gallon in 2015. Given the Services' goals, mid-range estimates suggest that DoD's drop-in renewable fuel use would represent an additional annual fuel cost of $865 million by 2015 and $2.2 billion by 2020, which represents a 10– 15 percent increase over just conventional petroleum fuels.

Although creating a new DoD commodity class for renewable fuels might help them gain greater leadership support and visibility, the associated redundant infrastructure costs and user acceptance impediments outweigh the benefits. DoD and Congress have other options to encourage the drop-in renewable fuels industry and help address production and cost barriers. DoD could enhance renewable fuel production through the use of long-term contracting authority or the Defense Production Act Title III Program. Congress could require the addition of renewable jet fuels to RFS2 volumetric mandates and make them eligible for production incentives. Congress could also directly support drop-in renewable and cellulosic fuel processing infrastructure.

In conclusion, increased DoD renewable fuel use contributes to U.S. national security interests,[1] achieves Service energy security goals, and offers some limited military utility. However, the projected supply of drop-in renewable fuels will not be sufficient to meet anticipated DoD demand for renewable jet fuel products. Also, price premiums for drop-in renewable fuels and the budgetary implications associated with meeting renewable fuel goals may be considerable. Further action by DoD and Congress could help to promote renewable jet fuel production and address the price premiums necessary for the Services to achieve their renewable fuel goals.

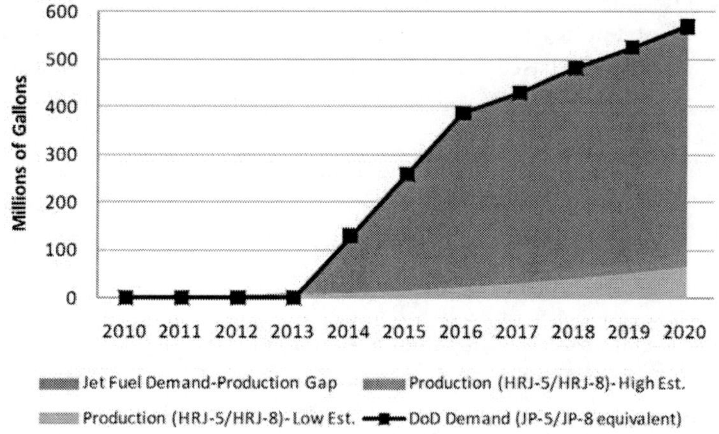

Figure ES-1. Comparison of DoD Demand for Renewable Jet Fuels and Projected Supply of These Drop-in Fuels, 2010–20.

1. INTRODUCTION

The Office of the Assistant Secretary of Defense for Operational Energy Plans and Programs and the Defense Logistics Agency, in conjunction with the military Services, prepared this report in response to Section 334 of the National Defense Authorization Act (NDAA) for Fiscal Year (FY) 2010. Section 334 requires the completion of two assessments focused on DoD's renewable fuel use:

1. An assessment of the use of renewable fuels, including domestically produced algae-based, biodiesel, and biomass-derived fuels, as alternative fuels in aviation, maritime, and ground transportation fleets (including tactical vehicles and applications). Such assessment shall include technical, logistical, and policy considerations.
2. An assessment of whether it would be beneficial to establish a renewable fuel commodity class that is distinct from petroleum-based products.

This report addresses and responds to the Congressional reporting requirements of NDAA FY10 Section 334. The assessment ultimately encompassed a complex combination of supply (industry technologies and capabilities) and demand (military Service requirements and plans) issues, coupled with a highly dynamic fuels market environment driven by political, statutory, and regulatory forces.

Objectives

Section 334 requires DoD to report on its current and potential use of renewable fuels, specifically in tactical systems. Any significant increase in renewable fuel use will likely impact DLA Energy's business processes as well as the entire petroleum operational community.

Therefore, our first objective was to understand DoD and Service use of existing renewable fuels in non-tactical vehicles (NTVs) and the potential tactical system and weapons platform (operational) use of next generation drop-in renewable fuels. We considered the technical, logistical, and policy constraints on the use of domestically produced ethanol, fatty acid methyl ester (FAME)-based biodiesel, biobutanol, biomass-derived or algae-based renewable gasoline, diesel, and jet fuels, such as cellulosic Fischer-Tropsch-derived and hydrotreated renewable fuels.

Our second objective was to examine the advantages (and disadvantages) of establishing a renewable fuel commodity class distinct from the petroleum-based products that DLA Energy currently procures on behalf of the Services.

To reach these objectives, we did the following:

- Collected and analyzed similar studies, completed and ongoing.
- Examined the potential for renewable fuels to support DoD requirements:
  - Analyzing the current state of renewable fuels, including the existing industrial base supporting the development of such fuels

- Gauging the production and distribution capabilities of the renewable fuel industrial base by category of product, including biomass-derived fuels
- Assessing the probable future (through 2020) of the renewable fuels production industrial base and distribution capabilities on the basis of technical, logistical, and policy implications
- Assessing whether the potential production capabilities could successfully integrate into the existing petroleum distribution infrastructure and the commercial changes and regulatory and policy actions needed so DLA Energy could procure and move these fuels to defense fuel support points.

- Assessed the potential for such biomass fuels to allow DoD—with DLA Energy as the supply chain executive agent—to comply with the Energy Independence and Security Act (EISA) of 2007 Section 526.
- Assessed whether establishing a renewable fuel commodity class distinct from petroleum-based products would be beneficial.

DoD Renewable Fuel Use and Potential Use

We address the supply side of the equation by examining 2009 production estimates of commercial renewable fuels and realistic projections of future renewable fuel production until 2020. We evaluate the current industrial base supporting the development of such fuels and the availability of feedstock.

Concurrently, we also examine the technical constraints on DoD use of renewable fuels, describing the anticipated DoD demand for these fuels, transportation and blending logistics, and implications of several management options. We describe these considerations and constraints in the context of existing statutes and policies. We summarize current energy goals for DoD fuel use and describe how each renewable fuel type can support them. We also offer interventions that could help develop a renewable fuel supply sufficient to meet the consumption goals of DoD's NTV fleet and Service tactical systems and weapons platforms.

Renewable Fuel Commodity Class Establishment

Using the results of the market, technical, and policy analyses, we discuss the policy changes needed to make renewable fuels more available and to manage these new fuels in the DLA Energy supply chain. We summarize the advantages (and disadvantages) of establishing a new renewable fuel commodity class, explore other tracking and management options, and highlight the policies or policy changes needed to pursue these options.

Scope

Renewable Fuels

Federal legislation, regulation, and policy recognize three major categories of non-petroleum liquid transportation fuels—alternative, renewable, and synthetic:

- Alternative fuels are transportation or mobility fuels not composed of or derived from liquid petroleum, including renewable and synthetic fuels. These fuels include petroleum liquid and alcohol blends containing 15 percent or less of petroleum that are pursuant to standard seasonal fuel specifications.

- Renewable fuels are transportation or mobility fuels, used alone or blended with petroleum-based fuel and wholly derived from biomass or its decay products. (This term can also refer to petroleum-blended fuel with a renewable component above a certain percentage of "neat" renewable fuel products, such as situations where B20 is termed biodiesel.)
- Synthetic fuels are liquid hydrocarbon fuels produced from coal, natural gas, or, increasingly, biomass.

These definitions are primarily derived from language found in the Energy Policy Act (EPAct) of 1992, EPAct 2005, and EISA 2007. (Table 1-1 shows the sources. Chapter 2 details the statutes.) We focus on fuels not derived from fossil fuels (alternative or synthetic fuels). However, synthetic fuels are increasingly drawn into this discussion when biomass feedstocks are combined with fossil-based synthetic crude feedstock for the synthesis of end-use fuels.

Per Section 334, we focus on renewable fuels—those that meet the definition of Section 211 of the Clean Air Act (CAA) of 1990, 42 *U.S. Code* 7545—which (1) are produced from renewable biomass; (2) are used to replace or reduce the quantity of fossil fuel present in a transportation fuel, heating oil, or jet fuel; and (3) have life-cycle greenhouse gas (GHG) emissions at least 20 percent less than conventional-fuel life-cycle GHG emissions.

Renewable biomass includes the following:

- Planted crops and crop residue harvested from existing agricultural land cleared or cultivated prior to December 19, 2007, which was non-forested and either actively managed or fallow on December 19, 2007
- Planted trees and tree residue from a tree plantation located on non-federal land that was cleared and actively managed prior to December 19, 2007
- Animal waste material and animal byproducts
- Slash and pre-commercial thinnings from non-federal forestland that is not ecologically sensitive
- Biomass (organic matter available on a renewable or recurring basis) obtained from the immediate vicinity of buildings and other areas regularly occupied by people, or of public infrastructure, in an area at risk of wildfire
- Algae
- Separated yard waste or food waste, including recycled cooking and trap grease.

Table 1-2 shows the renewable fuels that meet this definition.

Study Limits

First, pursuant to the language in Section 334, we primarily focus on U.S. domestic production of renewable fuels. Although international market fluctuations, treaties (such as Kyoto), and tariffs can indeed impact the domestic fuel markets, we focused our market analysis and projections on U.S. domestic renewable fuel production.

Table 1-1. Definitional Sources Surveyed

Fuel type	EPAct 1992, Sec 301(2)	26 *U.S. Code* 6426(d)(2)	42 *U.S. Code* 13211(2)	10 CFR 490.2	10 CFR 500.2	EPAct 2005	RFS1	EISA 2007	RFS2	EO 13423	EO 13514	Fleet Guidance for RFS2	DLA Energy
Alternative	✓	✓	✓	✓	✓	✓				✓		✓	
Renewable						✓	✓	✓	✓				
Synthetic					✓								✓

Note: CFR = *Code of Federal Regulations*; RFS1 = original Renewable Fuel Standard; RFS2 = expanded Renewable Fuel Standard; EO = Executive Order.

Table 1-2. Renewable Fuels

Fuel	Feedstock/production process
Ethanol	• Corn starch • Sugar cane • Wheat, barley, oats, rice, or sorghum • Non-cellulosic portions of separated food waste • Starches from agricultural residues or annual cover crops
Biodiesel Renewable diesel	• Soybean or canola oil • Oil from annual cover crops • Algal oil • Biogenic waste oils, fats, and greases • Non-food-grade corn oil
Cellulosic ethanol Cellulosic diesel Cellulosic naphtha Green gasoline Cellulosic jet fuel	• Agricultural residues, slash, forest thinnings and forest product residues, annual cover crops • Switchgrass and miscanthus • Cellulosic components of separated yard wastes • Cellulosic components of separated food wastes • Cellulosic components of separated municipal solid wastes (MSW)
Hydrotreated renewable jet fuel (HRJ)	• Camelina oil • Jatropha oil • Biogenic waste oils, fats, and greases (including tallow) • Soybean or canola oil • Algal oil
Biobutanol	• Corn starch • Algae
Hydrogen fuel	• Biomass • Algae

Second, given the inherent uncertainty in fuel markets and time scale of the Renewable Fuel Standard (RFS) 2 mandates, we limit our analysis and projections to the next decade (FY10–20). This limitation is consistent with the farthestreaching Services' goals for renewable fuel use in tactical systems and weapon platforms.

Table 1-3. Summary of Renewable Fuels Addressed

Fuel	Feedstock	NTV or operational fuel use	Chapter
Conventional Ethanol	Corn, other sugars, and starches	NTV only	5
Cellulosic Ethanol	General cellulose (energy crops, wood wastes, etc.)	NTV only	5
FAME Biodiesel	Vegetable oils or animal fats	NTV only	6
Hydrotreated Renewable Diesel	Plant oils (biomass-based)	NTV and operational use	6
Hydrotreated Renewable Jet	General triglycerides (animal fats, plant oils)	NTV and operational use	7
Cellulosic Diesel	General cellulose (energy crops, wood wastes, etc.)	NTV and operational use	6
Cellulosic Jet	General cellulose (energy crops, wood wastes, etc.)	NTV and operational use	7
FAME Biodiesel	Algal oils	NTV only	6
Hydrotreated Renewable Jet	Algal oils	NTV and operational use	7

Approach

This study consisted of three main research and analysis efforts: (1) literature review, (2) renewable fuel market analysis, and (3) statutory, regulatory, and policy analysis.

The statutory, regulatory, and policy analysis formed the basis of the report by taking a snapshot of the statutory, and environmental regulatory, frameworks currently in place (Chapters 2–4). DoD policies are summarized, and the historical and current use of renewable fuels addressed, as background for the fuel projections and goals.

For the literature review, we searched the web and reviewed relevant bibliographies for renewable fuel studies going back to about 2000, including relevant studies through April 2010. Our systematic approach included an investigation of DoD-specific gray literature studies, those by or for the federal government at large for non-military purposes, and open source, academic, or industry research performed outside the federal sector.

Using these resources, our comprehensive market analysis sought to determine the current production (including capacity utilization) of DoD relevant renewable fuels and to develop requisite resources to forecast their future production in FY10–20 (Chapters 5–7). This report addresses the DoD relevant fuels shown in Table 1-3.

Using this research, we evaluated the current industrial base supporting the development of fuel feedstock availability, processing, and transportation of finished product (summarized in Appendix A). This analysis was used to do the following:

- Project the overall production capacity of renewable fuels over the study time frame (FY10–20).
- Estimate the impacts of federal mandates on the production and availability of renewable fuels.
- Elaborate on the direct and indirect land use and other environmental constraints associated with production of renewable fuels.

- Project the availability of commercial infrastructure to supply each fuel type.
- Evaluate limitations throughout the supply chain, including production, storage, transportation, distribution, and commercial infrastructure.
- Determine transportation and blending logistics through the forecast period, including rail, pipeline shipping, and storage challenges, and refinery turnarounds.

The statutory, regulatory, and policy analysis also took the results of the market analysis to research the legislative and policy changes necessary to increase and ensure the availability of renewable fuels in DoD supply chains. Building on the statutory analysis, we identified and analyzed DoD and Service renewable fuel policies and goals to understand potential fuel demand and their implications in the context of the market study (Chapter 8). We engaged with key DoD, government, and industry stakeholders to identify and understand necessary policy interventions, enablers, and fuel management implications, including the establishment of a separate commodity class (Chapter 9).

2. DoD Historical Fuel Use

U.S. consumption of renewable energy represents 8 percent of the nation's total energy usage, with renewable fuels (biofuels, biogas, wood, and biomass waste) making up approximately 50 percent of that portion (Figure 2-1).

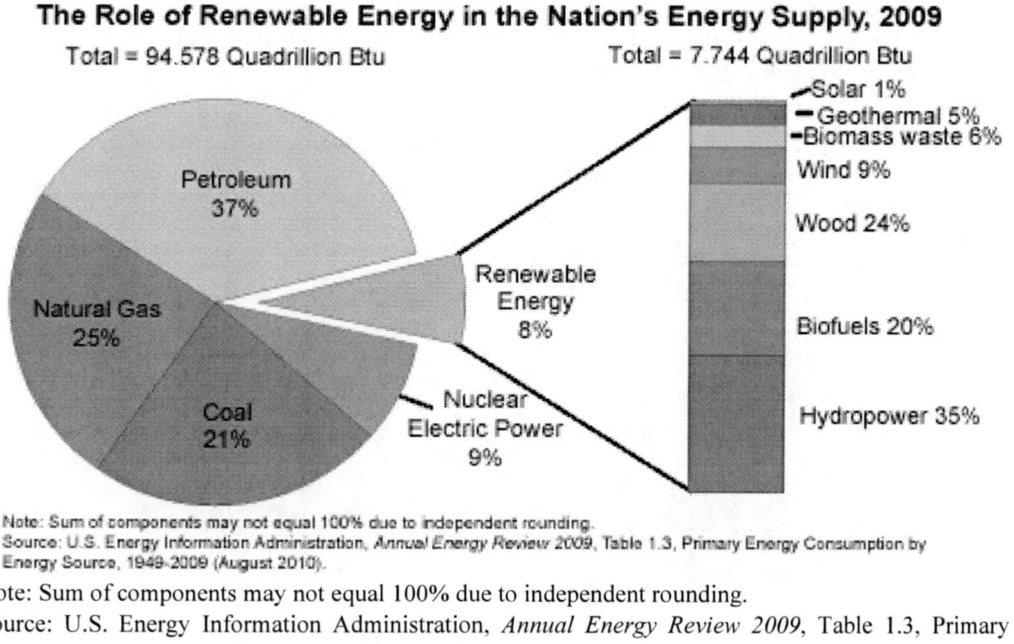

Note: Sum of components may not equal 100% due to independent rounding.

Source: U.S. Energy Information Administration, *Annual Energy Review 2009*, Table 1.3, Primary Energy Consumption by Energy Source, 1949-2009 (August 2010).

Figure 2-1. Renewable Energy Consumption.

Overall, U.S. consumption of biofuels has greatly increased over the past decade. Fuel ethanol increased from less than a billion gallons consumed in the early 1980s to more than10 billion gallons in 2009. Similarly, consumption of biodiesel in the early part of this decade was approximately 10 million gallons and has increased thirtyfold to more than 330 million gallons by 2009.

Dod Fuel Requirements and Use

DLA Energy purchased 128,965,000 barrels of petroleum in FY09.[2] The Energy Information Administration (EIA) reports DoD's total petroleum use at 679.7 trillion Btu, 92 percent of a total 737.2 trillion Btu used by the federal government (Figure 2-2).[3] DoD's use of petroleum is 1.9 percent of the total petroleum consumed by the United States in 2009 (35.268 quadrillion Btu).[4]

Jet fuel constituted 500.6 trillion Btu of the petroleum consumed by DoD, or 17.3 percent of the total used by the U.S. transportation sector in 2009 (2,889 trillion Btu).[5] Of the fuels reported by DLA Energy, approximately 115 million barrels (approximately 90 percent of the petroleum purchased) are attributed to use in tactical systems and weapons platforms.[6]

Dod Renewable Fuel Use

DoD consumption represents only a tiny fraction of total U.S. biofuels consumption. Table 2-1 shows comparison data for FY09, when DoD's use of B20 and E85 represented about 0.07 percent of the total U.S. consumption.

Table 2-1. DoD and U.S. Biofuel Consumption (thousands of GGEs)

Fiscal year	DoD consumption		Total U.S. consumption	
	Biodiesel (B20)	Ethanol (E85)	Biodiesel (B20)	Ethanol (E85)
2009	5,122	2,729	1,911,087	9,188,371

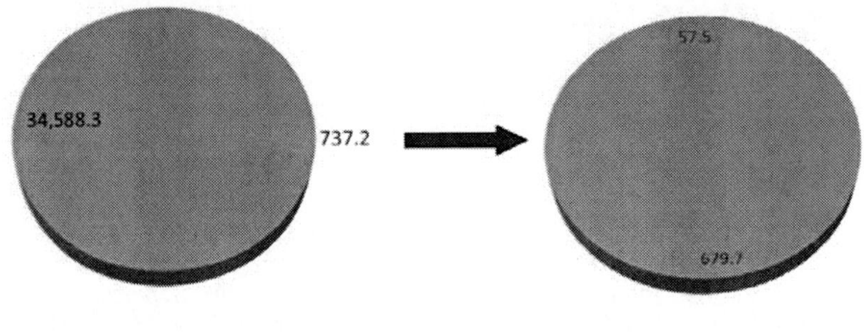

Source: EIA Annual Energy Review (AER) 2009.

Figure 2-2. DoD Petroleum Consumption.

NTV use accounts for almost all DoD biofuel consumption. DoD prohibits the use of biodiesel and E85 in tactical vehicles due to operational and mission-readiness concerns.

NTV Consumption

DoD consumption of renewable fuels has increased over the past 5 years. Table 2-2 illustrates the increase, about 25 percent, in gallons and gasoline gallon equivalents (GGEs). The increase is largely driven by a 500 percent increase in ethanol use.

Most of the NTVs that consume the biofuels (E85 and biodiesel) are E85 flex-fuel vehicles (FFVs) and diesel vehicles. Over the last 5 fiscal years, DoD efforts to comply with EPAct 1992 and Executive Order (EO) 13423 requirements are responsible for a significant increase in E85 FFVs and the use of biodiesel in DoD's NTV fleet. As a result, the percentage of gasoline-only vehicles in the NTV fleet decreased from 55.9 to 47.7 percent between FY05 and FY09.

Tactical System and Weapons Platform Consumption

To date, DoD tactical system and weapons platform renewable fuel consumption has been made up exclusively of test, qualification, and certification efforts. Initiatives underway as of fall 2010 include the following:

- Air Force test and certification efforts using 400,000 gallons of HRJ-8 from camelina and tallow derived fuels

Table 2-2. DoD Reported Use of Biofuels in NTVs, FY05–09 (thousands of GGEs or gallons)

Fiscal year	Biodiesel (B20)		Ethanol (E85)		Total	
	GGE	Gallon	GGE	Gallon	GGE	Gallon
2005	5,737	5,122	540	750	6,277	5,872
2006	6,060	5,400	800	1,112	6,860	6,512
2007	5,145	4,585	1,098	1,525	6,243	6,110
2008	5,333	4,748	1,390	1,930	6,723	6,678
2009	5,122	4,566	2,729	3,791	7,851	8,357

Source: Federal Automotive Statistical Tool (FAST) Database.

Table 2-3. Procurements of Renewable Operational Fuels

Use	Fuel type	Feedstock	Gallons
Aviation	HRJ-5	Algae-derived	1,500
	HRJ-5	Camelina	40,000
	HRJ-8	Camelina	161,350
	HRJ-8	Tallow	100,000
Diesel	HRF-76	Algae-derived	20,000
Total renewable fuel (neat product)			322,850

Source: DLA Energy, World Wide Energy Conference, *The Path to Commercially Viable Alternative Aviation Fuels* (presentation, May 12, 2010) and personal communication with Jeanne Binder, DLA Energy, and Scott Johnson, Sustainable Oils.

- Army test and qualification efforts for HRJ-8 using 34,950 gallons
- Navy test and certification using 190,000 gallons of HRJ-5 from camelina feedstock and 1,500 gallons of HRJ-5 from algae oils.

Table 2-3 shows purchases of renewable operational fuels.

3. STATUTES AND REGULATIONS

This chapter begins with the statutes and regulations that address alternative and renewable fuels, focusing on EISA. It elaborates on the specific GHG provisions included in EISA 526 and provides an overview of the Environmental Protection Agency (EPA) RFS. Finally, it covers renewable fuels and their environmental context for compliance with these regulations.

The major findings associated with this chapter:

- Expanded Renewable Fuel Standards (RFS2) excludes jet fuel, DoD's primary operational fuel, from volumetric production mandates.
- Few renewable jet fuel production pathways satisfy the life-cycle GHG requirements of RFS2 or Section 526 of EISA.

Statutory Framework

Alternative and renewable fuels are defined in EPAct and EISA and by EPA as part of the RFS.[7]

Alternative Fuels

Numerous statutes, regulations, and guidance contain the term *alternative fuel* (see Table 1-1). Most leverage the definition in EPAct 1992, Section 301(2), or cite it specifically:

> Methanol, denatured ethanol, and other alcohols; mixtures containing 85 percent or more (or such other percentage, but not less than 70 percent, as determined by the Secretary, by rule, to provide for requirements relating to cold start, safety, or vehicle functions) by volume of methanol, denatured ethanol, and other alcohols with gasoline or other fuels; natural gas; liquefied petroleum gas; hydrogen; coal-derived liquid fuels; fuels (other than alcohol) derived from biological materials; electricity (including electricity from solar energy); and any other fuel the Secretary determines, by rule, is substantially not petroleum and would yield substantial energy security benefits and substantial environmental benefits.

This definition establishes the primary distinction between alternative and renewable fuels: the alternative fuels category includes non-petroleum fossil fuels, while the renewable fuels category does not. Although these statutory definitions apply directly to NTV use, this report also applies them to use in tactical systems and weapon platforms, unless superseded by a DoD definition.

Renewable Fuels

The EPA RFS, included under subsection M of 40 *Code of Federal Regulations* (CFR) 80, provides the regulatory definition of *renewable fuel.* EPA has issued two versions of the standard. The first, RFS1, draws its language and authority from EPAct 2005. The second, RFS2, stems from EISA 2007 and supersedes RFS1. A comparison of the two standards highlights the evolution of the fuels considered renewable.

EPAct 2005 And RFS1

EPAct 2005 and RFS1 define renewable fuel as follows:

> (1) *Renewable fuel* is any motor vehicle fuel that is used to replace or reduce the quantity of fossil fuel present in a fuel mixture used to fuel a motor vehicle, and is produced from any of the following: (i) Grain. (ii) Starch. (iii) Oilseeds. (iv) Vegetable, animal, or fish materials including fats, greases, and oils. (v) Sugarcane. (vi) Sugar beets. (vii) Sugar components. (viii) Tobacco. (ix) Potatoes. (x) Other biomass. (xi) Natural gas produced from a biogas source, including a landfill, sewage waste treatment plant, feedlot, or other place where there is decaying organic material. (2) The term "Renewable fuel" includes cellulosic biomass ethanol, waste derived ethanol, biodiesel (mono-alkyl ester), non-ester renewable diesel, and blending components derived from renewable fuel.[8]

This definition includes all feedstocks from which renewable fuels can be derived. Unlike the RFS2 definition, it does not require lower GHG emissions than traditional (petroleum-based) fuels and is restricted to fuels that power motor vehicles.

EISA 2007 and RFS2

EISA 2007 and RFS2 define renewable fuel as follows:

> Renewable fuel means a fuel which meets all of the requirements of paragraph (1) of this definition: (1)(i) Fuel that is produced from renewable biomass. (ii) Fuel that is used to replace or reduce the quantity of fossil fuel present in a transportation fuel, heating oil, or jet fuel. (iii) Has lifecycle greenhouse gas emissions that are at least 20 percent less than baseline life-cycle greenhouse gas emissions, unless the fuel is exempt from this requirement pursuant to §80.1403. (2) Ethanol covered by this definition shall be denatured as required and defined in 27 CFR parts 19 through 21. Any volume of denaturant added to the undenatured ethanol by a producer or importer in excess of 2 volume percent shall not be included in the volume of ethanol for purposes of determining compliance with the requirements under this subpart.[9]

This definition requires renewable fuels to have lifetime GHG emissions 20 percent lower than a baseline (detailed in the following sections). It expands the terms to include all *transportation fuels,* defined as "fuel for use in motor vehicles, motor vehicle engines, nonroad vehicles, or nonroad engines (except fuel for use in ocean-going vessels),"[10] rather than just fuel for motor vehicles alone. Instead of defining acceptable feedstocks, it refers to them as *renewable biomass* and defines the term separately. This definition includes seven categories of biomass sources (feedstocks), which are further defined by conditions placed upon them. For instance, a planted crop such as switchgrass would only be considered renewable biomass under RFS2 if it were "harvested from existing agricultural land cleared or cultivated prior to December 19, 2007" (the day the standard went into effect). These conditions are another significant change from RFS1. Table 3-1 shows the RFS2 renewable biomass definition, organized by source category and condition.

Appendix E defines the terms used in the biomass definition—such as planted crops and ecologically sensitive—which are found in RFS2 in 40 CFR 80.1401.

EISA 2007 and RFS2 set emission thresholds as part of their definitions for renewable fuel and the three renewable fuel subcategories they establish (advanced biofuel, biomass-based biofuel, and cellulosic biofuel).[11] The thresholds are based on a percent reduction in life-cycle GHG emissions from the 2005 baseline GHG emissions of the gasoline or diesel fuel they replace. For renewable fuels, the reduction is 20 percent; for advanced biofuel, 50 percent; for biomass-based diesel, 50 percent; and for cellulosic biofuel, 60 percent. The calculation of the 2005 petroleum baseline is explained below.

Appendix A summarizes the renewable fuel production process, from feedstock input, through pre-processing and processing, to blending.

EISA Section 526

Section 526 of EISA 2007 provides:

> No Federal agency shall enter into a contract for procurement of an alternative or synthetic fuel, including a fuel produced from nonconventional petroleum sources, for any mobility-related use, other than for research or testing, unless the contract specifies that the life-cycle greenhouse gas emissions associated with the production and combustion of the fuel supplied under the contract must, on an ongoing basis, be less than or equal to such emissions from the equivalent conventional fuel produced from conventional petroleum sources.[12]

While this statute has not prevented DoD from meeting its missions needs, its interpretation has been disputed by some.

2005 Petroleum Baseline Model

EISA 2007 and RFS2 define renewable fuel, and three subcategories of renewable fuels, in part by meeting threshold levels for reduction of life-cycle GHG emissions from baseline 2005 gasoline or diesel life-cycle emissions.

EISA defines baseline life-cycle GHG emissions as "the average life-cycle greenhouse gas emissions ... for gasoline or diesel (whichever is being replaced by the renewable fuel) sold or distributed as transportation fuel in 2005."[13] EPA interprets this definition as including both foreign and domestically produced fuels sold in the United States in 2005, but not domestically produced fuels destined for export.

The RFS2 final rule uses the National Energy Technology Laboratory (NETL) Petroleum-Based Fuels Life Cycle Greenhouse Gas Analysis—2005 Baseline Model (2009) to determine baseline values.[14] The NETL model's baseline for lifecycle GHG emissions is based on EISA definitions and covers conventional petroleum-based fuels sold or distributed in the United States in 2005.[15] These fuels include those derived from coal, coal and biomass, and unconventional fossil energy resources. It utilizes countryand U.S.-site-specific values and is available on line.[16] An Air Force Research Laboratory (AFRL) study found this analysis the best current baseline estimate for EISA.[17]

Table 3-1. RFS2 Biomass Definition, by Source Category and Condition

Source	Condition
Planted crops and crop residue	Harvested from existing agricultural land cleared or cultivated prior to December 19, 2007, and that was nonforested and either actively managed or fallow on December 19, 2007
Planted trees and tree residue	From a tree plantation located on non-federal land (including land belonging to an Indian tribe or an Indian individual held in trust by the United States or subject to a restriction against alienation imposed by the United States) that was cleared at any time prior to December 19, 2007, and actively managed on December 19, 2007
Animal waste material and animal byproducts	
Slash and pre-commercial thinnings	From non-federal forestland (including forestland belonging to an Indian tribe or an Indian individual held in trust by the United States or subject to a restriction against alienation imposed by the United States) that is not ecologically sensitive forestland
Biomass	Obtained from the immediate vicinity of buildings and other areas regularly occupied by people, or of public infrastructure, in an area at risk of wildfire
Algae	
Separated yard waste or food waste, including recycled cooking and trap grease, and materials described in § 80.1426(f)(5)(i)	

Source: 40 CFR 80.1401.

Fuel Supply

EPA's Office of Transportation and Air Quality evaluated which renewable fuels meet the EISA 2007 life-cycle GHG emission thresholds, including raw material acquisition and transformation, fuel production, fuel transport, use, and end of life. Using its thresholds, EPA maintains that the following fuels meet the threshold reductions from the baseline:

- Ethanol from corn starch at new natural gas, biomass, or biogas fired facilities with advanced technologies
- Biobutanol from corn starch
- Biodiesel and renewable diesel from soy oil or waste oils, fats, and greases or from algal oils
- Ethanol from sugar cane, cellulosic ethanol, and cellulosic diesel.[18]

EPA's process included the review of studies, models, and satellite data on land conversion. EPA also noted that fuels from crop residues, forest materials after product production, secondary crops, separated food and yard waste, and perennial grasses meet threshold compliance determinations. It is modeling grain sorghum ethanol, woody pulp ethanol, and palm oil. Producers will need to prove that their fuels meet these thresholds. EPA is establishing a process for producers to petition for including other fuels under EISA.

The AFRL study found that in 2008 the life-cycle assessment of fuels was still substantially uncertain—in excess of 10 percent.[19] EPA continues to evaluate these uncertainties.

Fuels produced by the Fischer-Tropsch process, with current methods, generate more GHG emissions than standard petroleum production of fuels,[20] though the comparison is not precise, and the extent of excess GHG emissions depends on which of a variety of Fischer-Tropsch cycles is used. EPA and the Department of Energy (DOE) estimated the life-cycle GHG emissions for coal-to-liquids (CTL) synfuels,[21] finding that they exceed those of baseline 2005 gasoline or diesel lifecycle emissions. Fischer-Tropsch processes that use biomass as feedstock may have life-cycle GHG emissions lower than the EPA baseline 2005 gasoline or diesel life-cycle emissions, and one promising Fischer-Tropsch process mixes biomass and coal as feedstock.

Total life-cycle GHG emissions of fossil and Fischer-Tropsch fuels can be reduced significantly through carbon capture and storage (CCS). Current CCS technology adds significant energy cost to all fuel production methods, and no CCS technology is commercially competitive.

RFS Program

EPAct 2005 amended the CAA to establish the RFS program,[22] requiring a minimum volume of renewable transportation fuel produced each year. EISA 2007 revised these requirements, establishing new "volume standards for cellulosic biofuel, biomass-based diesel, advanced biofuel, and total renewable fuel that must be used in transportation fuel each year."[23]

Requirements

Under the RFS program, EPA "must set a standard for each year representing the amount of renewable fuel that each refiner, blender, or importer must use." The initial RFS program (RFS1) under EPAct 2005 was promulgated on May 1, 2007,[24] and established a maximum mandated volume of 7.5 billion gallons of renewable fuel by 2012. Under RFS1, standards were "expressed as a percentage of gasoline sold or introduced into commerce." Table 3-2 shows the targets specified under RFS1.[25]

Table 3-2. RFS1 Minimum Renewable Fuel Requirements (billion gallons)

Year	Renewable fuel requirement
2006	4.0
2007	4.7
2008	5.4
2009	6.1
2010	6.8
2011	7.4
2012	7.5

EISA revised the statutory requirements, increasing the 2012 renewable fuel mandate to 15.2 billion gallons. The new program, RFS2, extended the renewable fuel mandates through 2022, when the required volume will be 36 billion gallons.

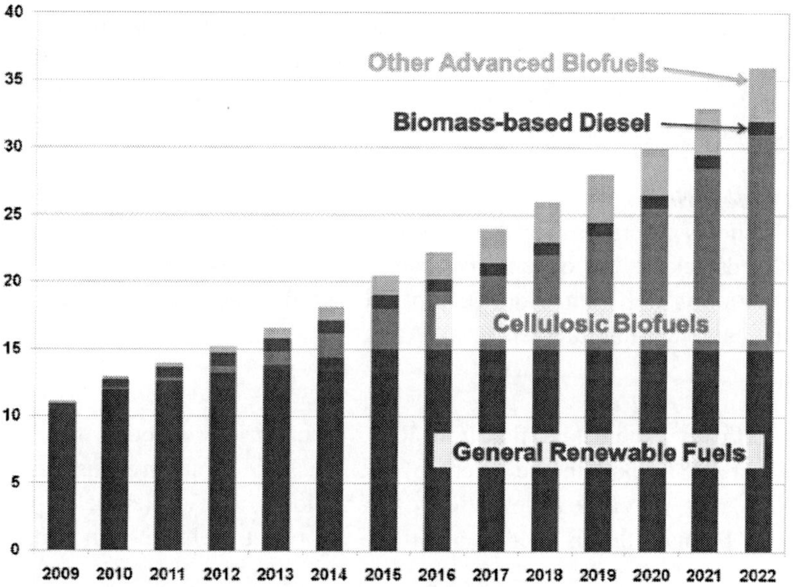

Note: Fuel quantities represent “neat,” rather than blended, fuel.

Figure 3-1. RFS2 Minimum Renewable Fuel Requirements (billion gallons).

Table 3-3. RFS2 Minimum Renewable Fuel Requirements (billion gallons)

Year	Total	Cellulosic biofuel	Biomass-based diesel	Total advanced biofuel[a]
2009	11.1	—	0.5	0.6
2010	12.95	0.1	0.65	0.95
2011	13.95	0.25	0.80	1.35
2012	15.2	0.5	1.0	2.0
2013	16.55	1.0	—[b]	2.75
2014	18.15	1.75	—[b]	3.75
2015	20.5	3.0	—[b]	5.5
2016	22.25	4.25	—[b]	7.25
2017	24.0	5.5	—[b]	9.0
2018	26.0	7.0	—[b]	11.0
2019	28.0	8.5	—[b]	13.0
2020	30.0	10.5	—[b]	15.0
2021	33.0	13.5	—[b]	18.0
2022	36.0	16.0	—[b]	21.0

[a] Total advanced biofuels include cellulosic biofuels, biomass-based diesel, and other unspecified advanced biofuels.

[b] To be determined by EPA through a future rulemaking, but no less than 1.0 billion gallons.

The RFS2 mandates include minimum volumes of four categories of renewable fuels: cellulosic biofuel, biomass-based diesel, advanced biofuel, and total renewable fuel. These renewable fuels are defined by production processes, feedstocks used to produce them, and new GHG emission thresholds as determined by lifecycle analysis."[26] As with RFS1, RFS2 standards are expressed as a percentage of fuel sold. However, the RFS2 standards are measured using the total gasoline and diesel used for on-road, off-road, locomotive, and domestic marine sectors ("MVNRLM"). Figure 3-1 and Table 3-3 show the minimum renewable fuel requirements established by the RFS2 for 2009–22.

RFS2 Fuel Categories

Table 3-4 shows definitions for the various categories of fuels. These definitions specify the types of feedstocks, types of land that can be used to grow and harvest feedstocks, and life-cycle reduction in GHG emissions (compared with the 2005 baseline average gasoline or diesel fuel that the renewable fuel replaces).

RFS2 Exclusion of Jet Fuel

EISA specifically excludes jet fuel from RFS2 volumetric production mandates. Sections 201 and 202 of EISA explain the intent of Congress in excluding this fuel. In Section 201(1)(A), Congress provides a definition of "additional renewable fuel"—fuel produced from renewable biomass that is used to offset the use of jet fuel or heating oil. This type of fuel is separate from "advanced biofuels," spelled out in Section 201(1)(B) and elaborated on in Table 3-4.

The regulatory provisions of RFS2, provided in Section 202(a)(1) of EISA, mandate that only transportation fuel used or sold in the United States meet volumetric increases in the use of renewable fuel, advanced biofuel, cellulosic biofuel, and biomass-based diesel. Because Congress did not require renewable fuel standards for "additional renewable fuel," jet fuel is excluded from the RFS2 production mandates. EPA addresses this exclusion in 40 CFR § 80.1407 by setting renewable fuel volume obligations for only gasoline and diesel fuels produced or imported into the United States.

Although the regulatory requirements omit jet fuel from RFS2 production mandates, producers or importers of renewable or cellulosic jet fuel can generate renewable identification numbers (RINs) if the fuel meets the definition of a renewable fuel in 40 CFR § 80.1401 and receives EPA approval. A fuel will qualify if it is produced from renewable biomass, replaces or reduces the quantity of fossil fuel present in jet fuel, and has life-cycle GHG emissions that are at least 20 percent less than baseline life-cycle GHG emissions. Certification by EPA for the renewable fuel pathway pursuant to 40 CFR § 80.1426 or § 80.1416 will result in the generation of RINs, which helps incentivize the jet fuel's production.

Environmental Regulatory Framework

The production of alternative and renewable fuels has varying impacts on air, water, and land use. Environmental regulations affect many of the activities involved in the siting, design, and construction of a renewable energy production facility. Renewable energy projects also must comply with a variety of applicable requirements during operation.

Producers must work within the environmental regulatory framework, and consumers may be required in the future to consider these factors in purchasing decisions.

Federal and state governments have instituted a range of statutes, regulations, and leadership-directed policies and goals aimed at reducing energy-related environmental impacts and increasing energy independence.[27] Although legislators, regulators, and policymakers have sought to encourage the development of renewable energy resources, they have generally done so within the established framework of existing laws and regulations. This includes regulatory and legal considerations of threatened and endangered species, sensitive habitats, cultural resources, land development constraints, water use restrictions and water pollution, and air pollution. In the subsections that follow, we discuss a few of the more likely regulatory influences on renewable energy production.

Table 3-4. RFS2 Renewable Fuel Categories

Category	Definition/feedstocks	Life-cycle GHG emission reduction (%)	Renewable fuels included
Type R—other renewable fuels	Any other transportation fuel not classified above produced from renewable biomass	20	Ethanol (corn starch) Ethanol (starches from agricultural residues or annual cover crops) Biobutanol (corn starch)
Type A—cellulosic biofuels	Any renewable fuel other than ethanol derived from cornstarch	50	Cellulosic ethanol Biomass-based diesel
Type B—biomass-based diesel	Biodiesel (defined in section 312(f) of EPAct 1992) produced from soybean oil; oil from annual cover crops; algal oil; biogenic waste oils, fats, and greases; non-food grade corn oil; or cellulose, hemi-cellulose, or lignin derived from renewable biomass	50	Biodiesel Renewable diesel Cellulosic diesel
Type C—cellulosic biofuels	Any cellulose, hemi-cellulose, or lignin derived from renewable biomass, including cellulosic biomass from agricultural residues, slash, forest thinnings, forest product residues, annual cover crops, switchgrass, and miscanthus; cellulosic components of separated yard wastes; cellulosic components of separated food wastes; or cellulosic components of separated MSW	60	Cellulosic ethanol Green gasoline Cellulosic diesel

Air Quality

The CAA limits certain pollutants in an effort to maintain minimum air quality standards throughout the United States. The set of regulated pollutants, known as “criteria pollutants,” include some released during fossil fuel combustion and energy production activities. The CAA regulates increases in emissions of these pollutants in geographic locations that do not meet the minimum air quality standards, which are known as non-attainment zones.

Fuels must meet the emissions standards for hydrocarbons, carbon monoxide, NO_x, and particulates set in the CAA. However, there are provisions where DoD can exempt certain vehicles (tactical systems and weapons platforms) from the CAA regulations for national security purposes.

The combustion of renewable fuels releases different amounts of pollutants than fossil fuels. Fuels with low sulfur content will reduce particulate emissions from aircraft gas turbines.[28] (Emissions of NO_x, SO_x, and other gases form particulate matter in the jet plume.) For example, biodiesel is generally low in sulfur, and therefore its particulate emissions are lower than those of regular diesel.

Renewable energy projects can provide regulatory relief in non-attainment zones. EPA has issued guidance that certain renewable energy projects generate emission reduction credits that can be used to comply with emission limits.[29] This latter benefit is of increasing interest given recent federal and state requirements to report and, in some cases, reduce GHGs.

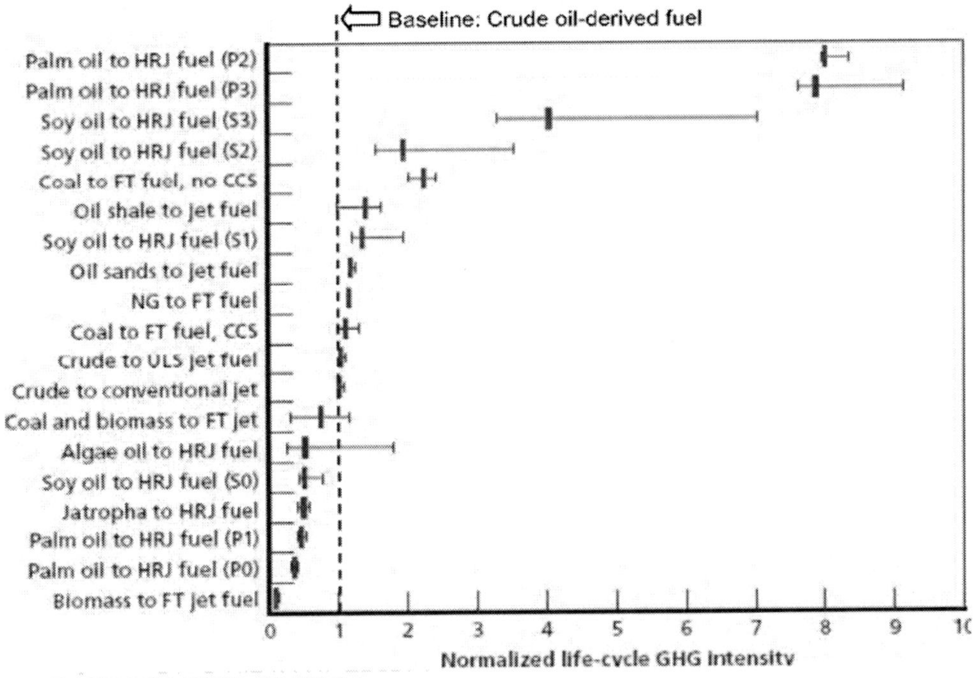

Source: RAND MIT (2009), TR554-8.1.

Figure 3-2. Normalized Life-Cycle GHG Emissions for Low, Baseline, and High Emission Cases for Jet-Fuel Pathways.

Greenhouse Gases

The U.S. Supreme Court (Massachusetts v. EPA) ruled in 2007 that air pollutants covered under the CAA do include GHGs. In the FY08 Consolidated Appropriations Act (Public Law 110–161), Congress directed EPA to publish a mandatory GHG reporting rule, which was published in the *Federal Register* in October 2009.[30] The EPA administrator released both an "endangerment finding" and a "cause or contribute finding" regarding GHG in December 2009, which has triggered new regulatory scrutiny that will likely impact energy generation and transportation systems.

EO 13514 reinforced the federal government's mandate to increase renewable energy use and specifically required federal agencies to account for and reduce their GHG emissions. This requirement is significant because renewable fuel use will likely be required for agencies to achieve federal GHG reduction goals. Pursuant to Section 2a, federal agencies, including DoD, submitted their GHG reduction goals. Those goals were aggregated to establish the new goal announced in January 2010, which requires the federal government to reduce its annual GHG by 28 percent by 2020.[31] Not only does this goal require aggressive energy efficiency and renewable source development, it now requires renewable projects to be evaluated in terms of their GHG emissions.

Life-cycle GHG emissions from renewable fuels depend on the full range of activities involved in producing the renewable fuel, including changes in land use required by the project. Although soil and plants sequester CO_2, these sequestered emissions are released when land is converted to cultivation. These emissions do not occur if the land itself was already cultivated and is simply repurposed. GHG emissions from land use changes can be direct, via the cultivation of the land, or indirect, by the cultivation of land leading to new land use in another location. Land use change can be a very large part of the total life-cycle GHG emissions of renewable fuel sources, such as corn.

Life-cycle GHG emissions vary greatly with fuel type. For example, fuels from oil sands used to derive Jet A would have greater life-cycle GHG emissions than conventional Jet A from a conventional mix of crude oils due to its requisite energy-intensive extraction and processing. GHG emissions from oil shale-derived fuel could be 50 percent higher than conventional Jet A.[32] Fischer-Tropsch jet fuels (FT-SynJet) from natural gas have emissions comparable to Jet A from oil sands, but FT-SynJet from coal has much higher GHG emissions than Jet A.[33] Figure 3-2 summarizes these data and shows life-cycle GHG emissions for production of jet fuel using a variety of fuel sources.

Considering the RFS2 standard, in which renewable fuels are those with life-cycle GHG emissions at least 20 percent lower than baseline life-cycle GHG emissions, only a few of these fuel sources qualify as "renewable." Most fuels produced with Fischer-Tropsch technology do not qualify as renewable under RFS2, but those using biomass do. However, when coal is mixed with biomass in a FischerTropsch process, the resultant fuel may qualify as renewable, depending on the proportion of biomass (as shown by the uncertainty range in Figure 3-2 for this fuel source). Algal-derived fuels vary greatly in life-cycle GHG emissions but could qualify as renewable, depending on CO_2 source, production method, and harvest process used.

Water Use

Regulations concerning water quality and quantity limit opportunities for locating renewable fuel feedstock and production facilities. Release of byproducts from renewable fuel projects near water bodies is restricted through the Safe Drinking Water Act or Clean Water Act (CWA). The CWA requires permits for polluted runoff from point sources as part of the National Pollutant Discharge Elimination System (NPDES).

In the production and refinement of alternative and renewable fuels, water regulations are a consideration in the technological feasibility and choice of location. Significant quantities of water can be required to grow biofuel and renewable fuel feedstocks and to subsequently convert or refine them into useable fuels, such as ethanol or biodiesels. Technology and regional circumstances impact the water requirements and resources available. The importance of these requirements depends on state and local water laws.

Conventional oil can consume large amounts of water as well as impact water quality in spills during transportation or extraction. Conventional fuel production uses water in large volumes for drilling, extraction, and conversion of crude oil and oil sands for alternative fuels.[34] According to an Argonne National Laboratory (ARNL) study, conventional petroleum production requires between 3.4 and 6.6 gallons of water per gallon of gasoline depending upon various factors, such oil well age, extraction approach, and water reuse. Further, 82 percent of oil sands in Canada are only recoverable via in situ technologies that require significant water and energy to create steam, which result in water intensities of 2.6 to 6.2 gallons of water per gallon of fuel produced.[35]

Renewable fuel feedstocks and production processes can likewise require significant amounts of water. Although fuels derived from forestry or crop residues; biogas from manure, whey, and paunch materials; and biodiesel from waste grease and oils do not always require significantly more water that petroleum fuels, several crops grown to produce fuel require substantial water, such as ethanol from corn, sugar cane, and switchgrass and biodiesel from rapeseed and other oils.[36] For instance, ARNL found that corn ethanol consumed 10 to 324 gallons of water per gallon of ethanol,[37] and switchgrass ethanol used 1.9 to 9.8 gallons of water per gallon, whereas conventional petroleum only ranges between 3.4 and 6.6 gallons of water per gallon.[38] Third generation renewable fuel systems, such as photosynthetic algae feedstock production, can require hundreds or even thousands of gallons of water per gallon of fuel produced.

Land Use

The feedstock requirements to produce the scale of renewable fuels required under RFS2 will represent a large investment of land area. Feedstock type and production pathways greatly vary in their land intensity. Although some cellulosic fuel systems could require monumental land area investments, feedstock production systems such as algae represent high-product-per-acre options that can lessen the required land footprint.

Producing feedstocks inherently involves land use, and whether or not the land has been cultivated before, the producer must comply with applicable federal, state, and local regulatory requirements in many broad categories:

- Threatened and endangered species
- Sensitive habitats or protected areas
- Cultural resources
- Land development constraints
- Aviation
- Telecommunications
- Air pollution
- Aesthetics

- Water use restrictions and water pollution
- Solid and hazardous waste
- Utility requirements.

Fuel feedstocks that can be grown on marginal land could represent a larger impact on biodiversity and sensitive habitats than feedstocks grown on land already cultivated. Further, crop cultivars chosen for their productive efficiency may also have implications as invasive species (new weeds, for example) that can outcompete native species and become potential nuisances.

4. DoD Policies, Organizations, and Efforts

This chapter summarizes alternative and renewable fuel-related DoD policies, organizations, and initiatives. It expands on the Services' policies, programs, and activities. The chapter ends with a discussion of projected goals based on current Service and DoD goals.

Directives, Instructions, and Regulations

The Office of the Secretary of Defense (OSD) issues DoD directives, instructions, and regulations for transportation policy,[39] which represent its direction and guidance. These policies outline the requirements for acquisition and procurement of all material and equipment related to transportation, including fleet (non-tactical) and tactical vehicles under *U.S. Code* Title 10.

DoD 4140.25-M, DoD Management of Bulk Petroleum Products, Natural Gas, and Coal

The objective of this policy is to demonstrate leadership in the acquisition and use of alternative fuels to lower national dependence on foreign petroleum products.

It notes that environmentally regulated fuels will be used to satisfy advanced emissions control systems used in administrative vehicles/equipment, as required by law or EO. Examples of regulated fuels include ultra-low sulfur diesel, blended gasoline such as E10 and E85, and biofuels. To the extent practical, no new combat support equipment or vehicles requiring gasoline are to be acquired or developed unless the support concept is to supply fuel as a package product.[40]

This policy also discusses general procedures for DLA Energy, DLA Energy field activities, defense fuel support points (DFSPs), other DoD components, and contracted DFSPs, to ensure uniform Business System Modernization-Energy (BSM-E) and Petroleum Quality Assurance training. Military Services must develop and conduct the Service training necessary to fulfill petroleum and energy management responsibilities.[41]

The policy describes DoD and federal agency policy and management responsibilities regarding the procurement, management, and use of biodiesel blend (B20) and fuel ethanol blend (E85) fuels in compliance with governing energy policy acts and EOs.[42]

In addition, Section C8.4 defines various alternative fuels as qualified fuels, including methanol, denatured alcohol, other alcohols, mixtures containing 85 percent or greater by volume of methanol, alcohols blended with gasoline and other fuels, compressed or liquefied natural gas, liquefied petroleum gas, hydrogen, coal-derived liquid fuels, fuels other than alcohol derived from biological materials (including blends containing 20 percent or greater by volume of biodiesel with diesel fuel), and electricity, including solar-generated electricity.[43,44]

DoD 4500.36-R, Management, Acquisition, and Use of Motor Vehicles

Chapter 14, "Use of Alternative Fuels and Vehicles," states that DoD encourages the use of all types of alternative fuels and alternative fuel vehicles (AFVs), as applicable, to support the National Strategy. It also states that DoD will acquire, through procurement or lease, alternative fueled and hybrid electric vehicles, use alternative fuels, reduce petroleum use in the agency's NTV fleet, and prescribe for the accounting of the vehicles, fuel types, miles driven, and operating costs.

It clarifies that the single fuel on the battlefield policy should not be used as a mechanism to dissuade the use of alternative fuels in administrative motor vehicles except when there is a compelling operational consideration, such as when commercial domestic suppliers cannot provide the alternative fuel or a national emergency exists.

Military Exchanges are encouraged to provide an alternative fuel or alternative fuels at locations where at least one of following can be supported: private-sector (public) vehicles, multiple federal agencies, the military family AFV, or an activity or installation fleet tailored to maximize Military Exchange support. DoD components are encouraged to develop and test innovative pollution prevention technologies at their facilities and installations.[45] Partnerships among industry, federal agencies, government laboratories, academia, and others are encouraged for the purpose of assessing and deploying innovative environmental technologies.

Section C14.2, "Acquiring Alternative Fueled and Hybrid Electric Vehicles," states that DoD components are required to meet the federal objectives for the acquisition of AFVs. The objectives may be met through the acquisition of light, medium, or heavy duty AFVs; through AFV credits offered for the use of dedicated alternative fueled light, medium, or heavy duty vehicles; or through the use of biodiesel fuel for up to 50 percent of the requirement.

The area where the AFV credit for acquisition, type of acquisition (dedicated and weight class), or biodiesel (B20) fuel use accrues is not restricted. For example, the acquisition of an AFV for use OCONUS will count as a credit toward the objective.

DoD components are encouraged to develop fueling infrastructure to meet the needs of the AFV fleet and strategy and to work with non-governmental commercial activities and with state and local entities to acquire alternative fuels at refueling facilities outside the installation. DoD components that have AFVs and a nearby source to refuel those vehicles must use alternative fuels in those vehicles.

Assistant Secretary of Defense for Operational Energy Plans and Programs

Section 902 of the NDAA FY09 added a new section 139b to Title 10, United States Code, to establish the position of the Director of Operational Energy Plans and Programs

(DOEPP) to "provide leadership and facilitate communication regarding, and conduct oversight to manage and be accountable for, operational energy plans and programs within the Department of Defense and the Army, Navy, Air Force, and Marine Corps." On January 10, 2011, the NDAA FY11 was signed into law and Section 901 (B) redesignated the DOEPP position to ASD (OEPP).

The ASD (OEPP) is the principal advisor to the Secretary and Deputy Secretary of Defense on operational energy security. By statute, the ASD (OEPP) is to

- provide leadership and facilitate communication regarding, and conduct oversight to manage and be accountable for, operational energy plans and programs within DoD and the Army, Navy, Air Force, and Marine Corps;
- establish the operational energy strategy;
- coordinate and oversee planning and program activities of DoD and the Army, Navy, Air Force, and Marine Corps related to
  - implementation of the operational energy strategy;
  - consideration of operational energy demands in defense planning, requirements, and acquisition processes;
  - research and development investments related to operational energy demand and supply technologies;
- monitor and review all DoD operational energy initiatives.

On June 25, 2010, Sharon Burke was sworn in as the first ASD (OEPP). Since this confirmation, the Office of the Assistant Secretary of Defense for Operational Energy Plans and Programs (OASD (OEPP)) has been established to "help the military Services and combatant commands improve military capabilities, cut costs, and lower operational and strategic risk through better energy accounting, planning, management, and innovation." OASD (OEPP) was expected to issue the *DoD Operational Energy Strategy* in December, 2010. Once released, it should set much of the future policy direction for DoD operation energy, including alternative and renewable energy.

DoD and DOE Memorandum of Understanding

A July 22, 2010, memorandum of understanding (MOU) between DOE and DoD covers efforts in the areas of energy efficiency, renewable energy, water efficiency, fossil fuels, alternative fuels, efficient transportation technologies and fueling infrastructure, grid security, smart grid, storage, waste to energy, basic science research, mobile/deployable power, small modular reactor nuclear energy, and related areas.

America's military pays a high price in terms of added cost, risk of life, and lost operational flexibility to deliver fuel supplies to forces in combat. Both agencies are committed to reducing these vulnerabilities through improved efficiencies and displacement of fossil fuels through on-site renewable power generation.

DoD aims to speed innovative energy and conservation technologies from laboratories to military end users, and it uses military installations as a test bed to demonstrate and create a market for innovative energy efficiency and renewable energy technologies.

DOE is the lead federal agency responsible for the development and deployment of advanced energy technologies, yet DoD will need to invest in many of these same energy technologies as well as other energy technologies unique to its operational requirements. Partnering with DoD gives DOE the opportunity to accelerate the deployment of its technologies and expertise to help address the critical economic and energy security needs of the United States and to promote scientific and technological innovation.[46]

Green Initiative for Fuels Transition

In 2009, the U.S. Pacific Command (PACOM), one of six regional combatant commands, initiated a "Green Initiative for Fuels Transition" (GIFTPAC) by forming an interagency working group, including representatives from DoD, U.S. Department of Agriculture (USDA), and DOE. The working group set three desired outcomes for renewable and alternative fuel in Hawaii:

1. Obtain a long-term contract from DLA Energy for a multiyear, stable-price supply of green fuel to replace 25 percent of DoD petroleum.
2. Incorporate agricultural, energy, environmental, government, industrial and commercial sustainability objectives into the enterprise model.
3. Reach an end state with sustainable competition for fuel in Hawaii.

DLA Energy purchases 78 million gallons of JP-8 jet fuel per year for Hawaii. To meet the PACOM 25 percent biofuel goal, it would need 20 million gallons of drop-in fuel. Estimates from producers found that 50 million gallons of algal oil will be needed—which would come from 25,000 acres of algae ponds. This will require large-scale development.

To reach these outcomes, the working group recommended an increase in contract duration beyond 5 years, a stabilization of pricing techniques, and collaboration with industry. In June 2010, DLA Energy published a request for information (RFI) to solicit information on sources of biofuels for Hawaii. The RFI specifically requires that the fuel be hydrotreated renewable jet (HRJ) or Fischer-Tropsch (FT-SynJet from biomass).

DLA Energy Alternative and Renewable Fuels Programs and Efforts

DLA Energy's mission is to "provide DoD and other government agencies with comprehensive energy solutions in the most effective and efficient manner possible." The DLA director's intent is that DLA be "positioned to enable and develop practical solutions that make our Nation less dependent on foreign sources of energy and compliant with environmental requirements."[47] To these ends, DLA Energy has three key business units that support DoD and military Service alternative and renewable fuel aspirations. First, the DLA Energy Research and Development program bridges the gap between proof of concept and commercial production capability. In doing so, it provides a pathway for technologies of value to DLA Energy and its customers to meet the wide-ranging needs of DoD. Second, the Energy Business Development Office is the lead agent for all DLA Energy business development efforts, including those related to alternative and renewable fuels. Finally, the

DLA Energy Plans and Programs Office integrates the organization's functions associated with bulk petroleum, alternative fuels, renewable energy, and energy conservation. Together, these business units directly focus on supporting the Services' alternative and renewable fuel goals and objectives and have already been successful in

- conducting seven research and development (R&D) studies to advance the state of knowledge involving the intricacies of alternative fuels and renewable energy development and the potential for operational usage;
- procuring 680,000 gallons of Fischer-Tropsch fuels;
- managing the Sustainment, Restoration and Modernization program that supports E85 and B20 biofuel goals; and
- procuring 600,000 gallons of biomass-derived hydrotreated renewable fuels in support of Services' certification and testing that has supported Air Force A-10 and Navy F/A-18 test flights.

In response to Services' renewable fuel requirements and requests, DLA Energy recently procured 1,500 gallons of algae-derived HRJ-5, 40,000 gallons of camelina HRJ-5, 20,000 gallons of algae-derived F76, 100,000 gallons of camelina HRJ-8, and 100,000 gallons tallow HRJ-8. It has also exercised options for 150,000 gallons of HRJ-5, 50,000 gallons of camelina HRJ-8, and 50,000 gallons of tallow HRJ-8. In addition, it has executed a contract for 34,950 gallons of HRJ-8.[48,49]

To support renewable supply assurance, DLA Energy also recently signed an agreement with the Air Transport Association (ATA) of America to promote widespread commercialization of environmentally friendly aviation fuels with less reliance on fossil fuels.[50] Representatives of DLA Energy and ATA signed a strategic alliance in Washington, DC, on March 19, 2010. The agreement highlights the shared goals of DoD and the principal U.S. airlines to advance the development and deployment of commercially viable and environmentally friendly alternative aviation fuels.

The alliance directs the formation of three collaborative teams, each of which will focus on specific developmental and marketing models of alternative fuels goals:

- The environment team will identify common methods for life-cycle assessment of GHG emissions for alternative aviation fuels.
- The deployment and logistics team will identify locations or regions suitable for alternative fuels production and deployment, as well as means of distribution to and from those locations.
- The contracting and finance team will jointly publicize supply opportunities, explore opportunities for complementary fuel-supply agreements, and develop compatible pricing and finance mechanisms.

DARPA Renewable Fuels Research

Since 2006, the Defense Advanced Research Projects Agency (DARPA) has been working to create military-grade jet fuel (JP-8) from renewable, cellulosic biomass and algal feedstocks. DARPA's efforts will help create fuel from a diverse source of feedstocks while expanding the technological base for producing JP-8 affordably.

DARPA's current biofuels program looks at cellulosic and algal feedstocks to demonstrate economic and scalable processes for producing JP-8 fuels from agricultural and aquacultural crops that will not compete with food supplies. DARPA seeks (1) to produce JP-8 that would cost less than $3 per gallon to produce, (2) a production cost at a moderate facility scale (less than 50 million gallons per year), and (3) to produce jet fuel that can be broadly and repetitively implemented to have pervasive impact on DoD. DARPA has found that jet fuel has been developed from algae using UOP hydro-cracking technology. This technology can convert algal oil, jatropha oil, and other agricultural crop oils into jet fuel. Several successful commercial test flights have used cellulosic and algal feedstocksderived jet fuel.

DARPA has several goals for its biofuels program goals for the cellulosic and algal feedstocks. The first goal for cellulosic-based jet fuel is to demonstrate 30 percent conversion efficiency, by energy content, of the feedstock material into the target JP-8 with a projected cost of production of JP-8 less than $3 per gallon at initial commercial-scale implementation (50 million gallons per year). The second goal for cellulosic-based jet fuel is to demonstrate 50 percent conversion efficiency, by energy content, with a projected cost of production of JP-8 less than $3 per gallon at initial commercial-scale implementation (50 million gallons per year).

DARPA's goal for algal-based jet fuel is a $2 per gallon triglyceride oil from algae with a projected cost of production of JP-8 less than $3 per gallon at an initial commercial-scale implementation (50 million gallons per year). The second goal for algal-based jet fuel is $1 per gallon triglyceride oil from algae with a projected cost of production of JP-8 less than $3 per gallon at an initial commercial-scale implementation (50 million gallons per year). DARPA plans first scalable demonstration testing in 2011 and a transition to full-scale production by 2013. DARPA currently has General Atomics and SAIC working on algal-based fuel, and Logos Technologies is currently working on the cellulosic material.

Service Policies and Programs

DoD Instruction 4170.11, *Installation Energy Management*,[51] provides procedures for DoD installation energy management and pertains to all phases of administration, planning, programming, budgeting, operations, maintenance, training, and material acquisition activities that impact the supply, reliability, and consumption of energy at DoD installations.

This instruction implements DoD Directive 4140.25, *DoD Management Policy for Energy Commodities and Related Services*, which updated the policies and responsibilities to (1) manage energy commodities, (2) minimize the number and complexity of fuels and maximize the use of commercial fuel, and (3) continue authorization to publish DoD 4140.25-M, *DoD Management of Bulk Petroleum Products, Natural Gas, and Coal* and other DoD publications on energy commodities.

As DoD's largest energy consumer, the Air Force was the first to pursue alternative fuels, and other Services followed suit. Now, all Service branches have published their own energy plans with goals and objectives on alternative fuels use.

Army

The Army released its *Army Energy Security Implementation Strategy* (AESIS) in 2009. The Army's strategy looks to increase energy security by forwarding energy options that ensure surety, survivability, supply, sufficiency, and sustainability. The AESIS's Energy Security Goal 3 is to "increase use of renewable and alternative energy."[52]

AESIS Objective 3.1 is to "substitute renewable resources for purchases of energy and fuel from fossil fuel sources where life-cycle is cost effective." The associated metric is stated as "% of electric and total energy from renewable sources."

AESIS Objective 3.2 is to "achieve the optimum mix of the most current vehicle technologies to reduce fossil fuel consumption and greenhouse gas emissions within the Army NTV fleet. (Such as Neighborhood Electric Vehicles [NEVs], Plug-In Hybrid Electric Vehicles, and Hydrogen Technologies)" by the end of FY12. This objective is primarily to meet federal requirements for use of non-tactical AFVs and to enable greater use of alternative fuels. The Army has not yet set specific renewable fuel use goal for tactical vehicles, but a future tactical fuel and energy strategy study is in process.

AESIS Objective 3.3 is to "transition from fossil fuel based tactical mobility/ power generation to alternative/renewable energy/sources." The metrics are driving much of the Army's current renewable qualification efforts. These metrics and targets are as follows:

- *Metric 3.3a.* Percentage of Army tactical ground equipment systems for which alternative or renewable fuels and synthetic fuel blend evaluations are completed. Targets are 50 percent by the end of FY12 and 100 percent by the end of FY14.
- *Metric 3.3b.* Percentage of Army engine and aviation systems for which alternative or renewable fuels and synthetic fuel blend evaluations are completed. Targets are 50 percent by the end of FY14 and 100 percent by the end of FY16.
- *Metric 3.3d.* Percentage of Army area of responsibility power generation requirements met by renewable or alternative sources. Target is 50 percent by the end of FY10.

In September 2010 the Army released the Tactical Fuel and Energy Implementation Plan. This plan, pending approval, recommends timelines, operational tasks, responsibilities and quantitative metrics to measure progress toward the AESIS Objectives and Metrics. For AESIS Energy Security Goal 3, the following quantitative metrics were proposed:

- By FY18, at least 15 percent of the training base's fuel requirement is met by alternative fuel blends.
- By FY23, at least 30 percent of the training base's fuel requirement is met by alternative fuel blends.
- By FY28, at least 50 percent of the training base's fuel requirement is met by alternative fuel blends.

Research, Development and Engineering Command (RDECOM) is the Army's technology leader and largest technology developer. Its mission is to "empower, unburden, and protect the Warfighter to enable the dominance of the Army," keeping the focus on the Soldier. RDECOM has eight major laboratories and research, development, and engineering centers, and three are the Army's focal points for alternative and renewable fuel efforts, which include the Tank and Automotive Research, Development and Engineering Center (TARDEC), Communications-Electronics Research, Development, and Engineering Center (CERDEC), and Aviation and Missile Research Development and Engineering Center (AMRDEC).

TARDEC develops, integrates, and sustains the Army's manned and unmanned ground systems and combat support systems. It has been involved with alternative fuels since the 1960s, played a key role in transitioning Army ground systems to the JP-8 single battlefield fuel in the 1980s, and was the early leader involved with the synthetic fuel testing and evaluation over the last decade. It is the key technical leader for the Army's efforts to qualify all tactical ground systems on 50/50 renewable fuel blends by 2014. In cooperation with other Services, TARDEC engages with the alternative fuels industry, performs benchtop fuel analysis (via the Army Petroleum Laboratory), engine evaluations, integrated systems evaluations, and demonstrations. It is currently responsible for developing and executing the Army's "Tactical Ground Equipment Qualification Roadmap" to qualify 50/50 blends of both Fischer-Tropsch synthetic paraffinic kerosene (FT-SPK) and HRJ with JP-8 for use in all Army ground systems by 2014.

CERDEC is responsible for research, development, test, and evaluation (RDT&E) of command, control, communications, computers, intelligence, surveillance and reconnaissance (C4ISR) technologies and has been collaborating with TARDEC on the use of 50/50 blends of both FT-SPK and HRJ with JP-8 in forward deployable mobile electric generators. Once these generator sets have been tested for performance, operability, and reliability on these synthetic and renewable fuels, the Program Manager-Mobile Electric Power (PM-MEP) will accept the results as qualification.[53]

AMRDEC is the focal point for providing research, development, and engineering technology and services for Army aviation systems.[54] It is responsible for qualifying all of Army aviation's weapon systems on synthetic and renewable fuels by 2016. In May 2010, the Army performed a flight demonstration test at Redstone Airfield and proved the viability of using a 50/50 blend of FT-SPK (coal to liquid) and JP-8 using the Black Hawk helicopter.[55,56] Currently, the Army is working to obtain Air Force certification for H-60 aircraft (Black Hawk to the Army and Pave Hawk to the Air Force) to fly on the 50/50 FT-SPK/JP-8 blend.[57]

Air Force

The Air Force 2010 *Energy Plan* sets energy end state goals for 2030. One of these goals is to fly aircraft on alternative fuel blends when they (1) are costeffective, (2) are domestically produced, and (3) have a life-cycle GHG footprint equal to or smaller than that of petroleum. The plan also establishes the Air Force goal to acquire 50 percent of the domestic aviation fuel requirement via an alternative fuel greener than conventional petroleum by 2016.[58]

The Air Force is already working to certify a 50/50 blend of JP-8 for use in aircraft. In its energy strategy released in 2009, it set goals to certify the entire fleet to use a 50/50 synthetic blend by 2011.[59]

To meet these goals, the Air Force alternative and renewable fuel effort is leveraging an integrated multi-organizational approach throughout the technology maturity spectrum. Beginning around 2005, this team and process was developed starting with the SPK fuel testing and certification efforts. On the basis of these initial processes and experience, three key organizations are currently responsible for the JP-8 specification (DoD's "single battlefield fuel") and for the testing, certification, and acceptance of HRJ (or bio-SPK). They are AFRL, the Alternative Fuels Certification Office (AFCO), and the Air Force Petroleum Agency (AFPA).

AFRL is the Air Force's R&D organization, which operates under the Air Force Materiel Command (AFMC) and manages the Air Force's science and technology program. The AFRL Propulsion Directorate is located at Wright-Patterson Air Force Base (AFB) and Edwards AFB with the mission "to create and transition propulsion and power technology for military dominance of air and space." It provides technical and testing of new fuels prior to certification testing.

AFCO was established in 2007 to manage certification of all Air Force platforms, support equipment, and base infrastructure for alternative JP-8 type fuels. Its first goal was to certify Air Force platforms for use of a 50/50 SPK fuel and JP-8 by early 2011. Using this Fischer-Tropsch experience, AFCO has more recently been asked to support the "concurrent" certification of these Air Force platforms for the use of HRJ.

AFPA is a field operating agency reporting to the Director, Logistics Readiness, Office of the Deputy Chief of Staff for Logistics, Installations, and Mission Support, Headquarters Air Force. It provides the Air Force with comprehensive information, services, and technical expertise needed for fuel support to air, space, and ground forces across the spectrum of operations. The agency provides a full range of technical and professional services related to fuels, propellants, chemicals, lubricants, gases, and cryogenics for all aerospace vehicles, systems, and equipment. The agency investigates alternative fuels, performs suspect product/ systems contamination investigations, handles laboratory analyses, and develops fuel quality assurance/surveillance standards and fuel product specifications. AFPA is also the Air Force's service control point for all DLA fuel-related support issues.

For example, the Air Force demonstrated success in March 2010 using blended HRJ-8 while testing an A-10C Thunderbolt II at Elgin AFB, FL. The jet was fueled with a blend of HRJ-8 and JP-8. The flight was conducted by members of the 40th Flight Test Squadron, a developmental test squadron that is part of the Air Armament Center. The week of the flight, ground tests were per-formed, and the A-10 flew with the fuels split into its two separate fuel tanks. (The A-10 has the ability to segregate its fuel system so one set of fuel tanks can be paired to one engine while the other set can be paired to the other without mixing fuel between systems, making it the perfect platform for testing fuel blends.)

Navy

The Navy released *A Navy Energy Vision for the 21st Century* in October 2010. This new strategy outlines the Navy's energy vision, which values energy as a strategic resource, understands energy security as fundamental to executing the Navy's mission afloat and ashore, and creates a Navy that is resilient to any potential energy future.

The Navy translates this vision into "strategic imperatives" (assure mobility, protect critical infrastructure, lighten the load, expand tactical reach, and green our footprint) that will facilitate the achievement of the Secretary of the Navy's five energy goals, which are as follows:

1. *Increase Alternatives Afloat.* By 2020, 50 percent of total Navy energy consumption will come from alternative sources.
2. *Increase Alternatives Ashore.* By 2020, the Navy will produce at least 50 percent of shore-based energy requirements from alternative sources.

3. *Sail the "Great Green Fleet."* The Navy will demonstrate a Green Strike Group in local operations by 2012 and sail it by 2016.
4. *Reduce Non-Tactical Petroleum Use.* By 2015, the Navy will reduce petroleum use in the commercial vehicle fleet by 50 percent.
5. *Energy Efficient Acquisition.* Evaluation of energy factors will be mandatory when awarding contracts for systems and buildings.

To achieve the third goal, the Navy will require 100,000 gallons of neat HRJ-5 and 350,000 gallons HRF-76 fuels. These renewable fuels will need to be blended with equal volumes of JP-5 and F-76, respectively. These 50/50 blends of HRJ-5 and HRF-76 will be used to sail the "Green Strike Group" in 2012. The Navy then plans to sail the "Great Green Fleet" in 2016 using 50/50 blends. A carrier task force is anticipated to require the use of 80,000 barrels of renewable fuel (1,680,000 gallons of neat HRJ-5 and 1,680,000 gallons of neat HRF-76).[60]

The *Energy Program for Security and Independence* sets the Navy's energy course to achieve its energy goals through five strategic program elements focused on increasing the energy efficiency of tactical and shore systems, increasing the use of alternative energy, and maintaining a steadfast commitment to environmental stewardship. These strategic program elements are as follows:

- *Energy efficient acquisition.* Incorporate energy efficiency into decisions for new systems and buildings.
- *Energy management.* Create an energy management structure through improved governance, planning, programming, and budgeting.
- *Science and technology.* Accelerate the adoption of nascent advanced technologies and foster partnerships with other federal agencies, universities, and laboratories to research, mature, and demonstrate new technologies.
- *Behavior change.* Improve energy management communication and awareness through training, education, and recognition programs to drive culture change.
- *Strategic partnerships.* Cultivate partnerships with existing and new organizations and leverage partner resources and capabilities to construct and execute the Navy's energy programs.

As evidence of the Navy's commitment to its energy security goals, on April 22 it conducted a test flight of an F/A-18 Super Hornet multirole fighter jet, nicknamed the "Green Hornet," powered by a 50/50 blend of conventional jet fuel and a biofuel that comes from camelina. "The alternative fuels test program is a significant milestone in the certification and ultimate operational use of biofuels by the Navy and Marine Corps," said Secretary of the Navy, Ray Mabus. "It's important to emphasize, especially on Earth Day, the Navy's commitment to reducing dependence on foreign oil as well as safeguarding our environment. Our Navy, alongside industry, the other Services and federal agency partners, will continue to be an early adopter of alternative energy sources." The Green Hornet biofuel program is the first aviation test program to test and evaluate the performance of a 50/50 biofuel blend in supersonic (above mach 1) operations—a critical test point to successfully clear the F/A-18 E/F for biofuel operations through its entire flight envelope. Once successfully demonstrated on the F/A-18 F414 engine, the Navy will expand its certification efforts to other Navy and

Marine Corps aircraft and Navy tactical systems. The Navy's ultimate goal is to develop protocols to certify renewable fuels for use in its aircraft and ships.[61]

Marine Corps

In August 2009, the Commandant declared energy a top priority for the U.S. Marine Corps (USMC).[62] In October 2009, the Commandant created the USMC Expeditionary Energy Office (E^2O), with the mission of analyzing, developing, and directing "the Marine Corps' energy strategy in order to optimize expeditionary capabilities across all warfighting functions."[63] The Marine Corps is finalizing a new *Expeditionary Energy Strategy* and *Implementation Guidance*. The two documents align the Marine Corps with guidance and mandates for operational and installation energy established by civilian and military leadership.

The Marine Corps consumes approximately 200,000 gallons of fuel per day in Afghanistan. A 60 kW generator fueled by an 80/20 mix of JP-8/cotton oil is being tested, as part of a small pilot program aimed at assisting the Afghans to use biofuels.

For vehicles operated at its bases, the Marine Corps has reduced fuel consumption by using B20—a mixture of 20 percent biodiesel made from domestically produced virgin soybeans and 80 percent diesel. In 2002, the Marine Corp implemented an educational campaign about B20. It demonstrated that B20 can be used in petroleum diesel vehicles without interruption and without modifications to the vehicle fuel systems or engines. The Marine Corps then uniformly converted fuel storage by cleaning out the existing diesel tanks and replacing the fuel with virgin soy-based B20. When using B20, the engine has only a 1 to 2 percent loss in power, torque, and fuel efficiency.

Although biodiesel is used in NTVs, Tri-Service Petroleum Oil Lubricant User Group policy currently forbids its use in tactical vehicles.

Anticipated DOD and Service Goals

The Services are setting goals for using renewable fuels in their tactical systems— as well as meeting the goals set forth in EOs for NTVs. As described above, the Army, Air Force and Navy set goals for how much alternative fuel they will purchase.

When projected to 2020, total demand from DoD to meet these goals for operational fuels grows from 0.59 million gallons of neat fuel product in 2010 to 745.97 million gallons in 2020 (Figure 4-1).

Operational fuel demand of the military Services comprises the renewable equivalents of F-76 (marine diesel) and JP-5/8 (jet fuel). The demand for jet fuels begins to increase in FY13 and constitutes more than 76 percent of the total demand for renewable operational fuels by 2020.

As part of their energy strategies, each Service is planning to certify the use of alternative fuel blends with their tactical systems to meet their consumption goals. These fuels must be able to be "dropped in" to current systems and meet standards for energy density, flash point, freezing point, thermal stability, lubricity, and viscosity.

Chapters 5 through 7 present a market analysis of renewable fuels by type. Chapter 8 presents a discussion on the technical feasibility of adopting different fuels for tactical purposes and presents an analysis of how these fuels can meet the Services' goals.

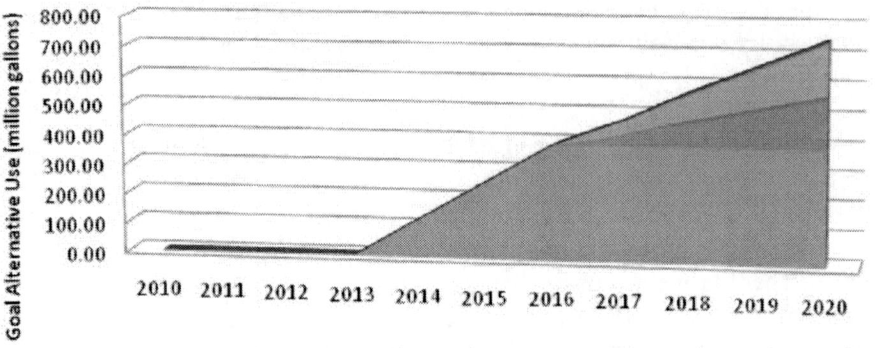

	2010	2011	2012	2013	2014	2015	2016	2017	2018	2019	2020
Army Diesel/F-76	0.00	0.00	0.00	0.00	0.00	0.00	0.00	0.00	5.55	6.66	7.77
Navy F-76	0.02	0.15	0.35	0.35	0.28	0.56	0.84	42.00	84.00	126.00	168.00
Army JP-5/JP-8	0.02	0.02	0.00	0.00	0.00	0.00	0.00	0.00	10.50	12.60	14.70
Navy JP-5/JP-8	0.04	0.15	0.10	0.10	0.28	0.56	0.84	42.00	84.00	126.00	168.00
Air Force JP-5/JP-8	0.40	0.40	0.40	0.40	129.43	258.47	387.50	387.50	387.50	387.50	387.50

Note: The Services' goals extrapolated fuel quantities represent neat, rather than blended, fuel.

Figure 4-1. Renewable Operational Fuel Demand (by Service), 2010–20 (million gallons).

5. Ethanol Market Projections

In this chapter, we present our assessment of the current and future commercial availability of ethanol. Conventional or corn-based ethanol currently represents the largest renewable fuel in the United States, with more than 10 billion gallons produced in 2009. Most of this ethanol (99.5 percent) is sold as "gasohol" or "E10" (typically, a blend of 90 percent gasoline and 10 percent ethanol),[64] and the remainder is sold as E85 (typically, 85 percent or more denatured ethanol blended with gasoline). Currently, other ethanol feedstocks represent only a small portion of production in the United States.

The market for ethanol is forecast to increase dramatically over the next 10 years, driven by the federal RFS1 and RFS2 mandates, emergence of cellulosic ethanol, and chemical properties that make ethanol a relatively good gasoline additive as an oxygenate and octane enhancer. We forecast that almost all (more than 99.3 percent) ethanol will be blended as E10 or E15 until the fuel additive market is saturated in 2015. The projected ethanol production will likely exceed both the E10 and E15 markets, necessitating substantial increased blending and sales of E85. In 2016 and onward, we expect the E85 market to grow significantly as almost all increases in ethanol production are blended as E85. The major hurdle to using E85 is commercial availability, which is limited and highly regionalized. Over the next 5 years, fueling infrastructure will develop to support projected E85 sales. In this chapter, we discuss the projected commercial availability of ethanol blended as E10, E15, and E85 from 2010 through 2020.

We forecast that national ethanol retail sales will increase more than 73 percent to 18,660 million gallons by 2020 (Table 5-1). In the short term (through 2013), the primary factor limiting ethanol availability is availability of corn feedstock. As production capacity for cellulosic ethanol comes on line after 2013, cellulosic production capacity will be the bottleneck in ethanol production.

To forecast commercial availability of conventional and cellulosic ethanol, we consider each supply chain stage, including feedstock availability, fuel production (capacity and projected production), transportation of fuel (to bulk facilities for blending or to a retail facility), and retail sales. Market conditions, costs, and bottlenecks at each of these stages of the supply chain determine the availability of biofuels. Table 5-1 summarizes ethanol forecasts for all supply chain stages from 2009 to 2020, the details of which are provided in the following text.

Ethanol Supply Chain Overview

Ethanol, or ethyl alcohol, is produced through the fermentation and distillation of simple sugars. Ethanol can be made from a wide array of biological feedstocks that contain either substantial amounts of sugar or materials that can be converted into sugar (such as starch or cellulose). In the United States, most ethanol is currently produced from corn. Other potential feedstocks include grasses (cellulose) and sugars, but their production volumes are limited due to cost and feedstock availability. Cellulosic ethanol is expected to emerge as renewable fuel within the next few years due to both mandates and improved commercial viability.

As of January 2010, 189 ethanol refineries were operating in the United States, at a total annual production capacity of 11.9 billion gallons (4.9 billion gallons in 2006).[65] An additional 16 biorefineries are being built or expanded that will add 1.4 billion gallons of new capacity.[66] Coupled with idle capacity of 1.2 billion gallons, the total ethanol annual production capacity is expected to increase to 14.5 billion gallons within the next few years. Almost all of these refineries produce ethanol from corn, so they are concentrated near the feedstock source, the Corn Belt in the Midwest.

Denatured ethanol is transported from the refinery to either a bulk terminal or a redistribution bulk terminal (Figure 5-1). Most ethanol is transported to terminals on the freight rail system, and the remainder is transported via tanker truck or barge.[67] From the terminals, ethanol is transported (typically after blending with gasoline) to retail locations by tanker truck for sale to end users. Most ethanol 0.5 percent) is sold as E10,[68] and the remainder is sold as E85.

Ethanol Feedstocks

The forecast for feedstocks for ethanol production through 2020 is as follows:

- *Corn will continue as the feedstock for roughly 98 percent of firstgeneration ethanol produced.* Corn will remain the preferred feedstock for ethanol production due to its availability, commodity cost, and economics for ethanol production. Corn is the only first-generation feedstock that enables ethanol to compete with the cost of petroleum at the scale necessary to meet demand (billions of gallons annually).

Table 5-1. Biodiesel and Renewable and Cellulosic Diesel Forecasts for Supply Chain Stages, 2009–20 (million gallons)

Stage	Type	2009[a]	2010	2011	2012	2013	2014	2015	2016	2017	2018	2019	2020
Feedstock availability	*Conventional Ethanol*												
	Corn[b]	10,606	11,810	12,430	12,910	13,395	13,870	14,140	14,400	14,600	14,800	15,000	15,075
	Others[c]	246	248	278	297	307	315	329	341	346	350	352	355
	Total	10,852	12,058	12,708	13,207	13,702	14,185	14,469	14,741	14,946	15,150	15,352	15,430
	Cellulosic Ethanol												
	Cellulose[d]	33.0 to 99.4 Billion Gallons											
	Total	10,852	12,058	12,708	13,207	13,702	14,185	14,469	14,741	14,946	15,150	15,352	15,430
Production capacity	*Conventional Ethanol*												
	Corne	11,649	12,251	12,853	13,455	14,057	14,293	14,529	14,764	15,000	>15,000	>15,000	>15,000
	Otherse	225	248	278	297	307	315	329	341	346	350	352	355
	Total	11,874	12,499	13,131	13,752	14,364	14,608	14,858	15,105	15,346	>15,350	>15,352	>15,355
	Cellulosic Ethanol												
	Cellulose	3	3	8	167	344	600	1,022	1,456	1,878	2,389	2,900	3,589
	Total		12,502	13,139	13,919	14,708	15,208	15,880	16,561	17,224	17,739	18,252	18,944
Projected production	*Conventional Ethanol*												
	Corn	10,606	11,810	12,430	12,910	13,395	13,870	14,140	14,400	14,600	14,800	15,000	15,075
	Others	150	248	278	297	307	315	329	341	346	350	352	355
	Total	10,756	12,058	12,708	13,207	13,702	14,185	14,469	14,741	14,946	15,150	15,352	15,430
	Cellulosic Ethanol												
	Cellulose[f]	2	5	7.3	150	310	540	920	1,310	1,690	2,150	2,610	3,230
	Total		12,063	12,715	13,358	14,012	14,727	15,387	16,057	16,639	17,301	17,962	18,660
Retail sales	E85[d]	54	58	61	69	76	81	101	1,139	1,486	1,844	2,431	3,229
	E10 or E15	10,702	12,005	12,654	13,289	13,936	14,646	15,286	14,918	15,153	15,457	15,531	15,431
	Total[d,e,f]	10,756	12,063	12,715	13,358	14,012	14,727	15,387	16,057	16,639	17,301	17,962	18,660

[a] 2009 values represent actual, not projected, data.

[b] U.S. Department of Agriculture (USDA), *USDA Agricultural Projections to 2019,* OCE-2010-1, February 2010.
[c] Food and Agricultural Policy Research Institute (FAPRI) at The University of Missouri, *U.S. and World Agricultural Outlook,* 2010.
[d] EIA, *Annual Energy Outlook 2010*, DOE/EIA-0383(2010), May 11, 2010.
[e] Renewable Fuels Association, *2010 Ethanol Industry Outlook,* February 2010, www.ethanolrfa.org/pages/annual-industry-outlook.
[f] EPA, *Renewable Fuel Standard Program (RFS2) Regulatory Impact Analysis*, EPA-420-R-10-006, February 2010.

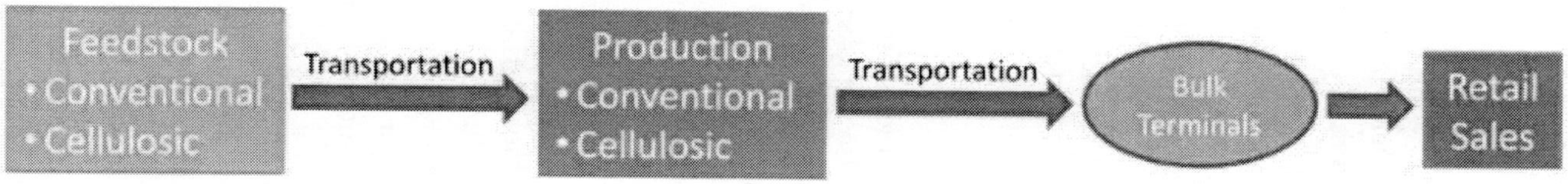

Figure 5-1. Ethanol Supply Chain.

- *Corn ethanol production will reach the limit of corn availability (15 billion gallons) in 2019.* The theoretical limit of corn available for conversion into ethanol is 15 billion gallons. We project corn ethanol production to meet this limit in 2019, and remain near that level over the long term.
- *Sugar will not significantly contribute to the ethanol market.* Sugar is grown in only four U.S. states and Puerto Rico, so it is not available on the scale necessary to produce significant quantities of ethanol. The availability of sugar as a feedstock for ethanol production is further limited by its importance in the food market—diverting its use to ethanol would compete with its use as a food product. In addition, although sugar cane has a higher yield than corn for ethanol production, its domestic commodity cost leads to a much higher ethanol cost per gallon than corn.
- *The cellulosic pathway will achieve commercial viability around 2012 and support most ethanol production growth afterwards.* Second-generation ethanol production from cellulosic biomass emits less GHG, has an abundance of available feedstock, and costs far less than corn-based ethanol production. EPA and DOE project commercial demonstration of a viable pathway by 2012 and commercially viable ethanol production from cellulosic biomass beyond that date.

First Generation

Almost all of the ethanol produced today is considered a first-generation biofuel, or a biofuel generated from crops or animal fats that can also be used as food and feed. The primary potential first-generation feedstocks for ethanol production in the United States through 2020 are as follows:

- *Corn.* Although estimates vary, roughly 98 percent of the ethanol produced in the United States comes from corn.
- *Sugars and starches (first-generation biofuel feedstock).* Sugar-based feedstocks include sugar beets, sugar cane, potato starches, and sweet sorghum. Ethanol production using sugar-based feedstocks is more efficient because the initial enzymatic step required for corn processing is unnecessary. However, only one refinery in the United States produces ethanol from sugar, and its annual capacity is only 1.5 million gallons. Growing conditions (sugar beets must be rotated with other crops) and government tariffs are primarily responsible for limiting sugar as a feedstock in the United States.
- *Other feedstocks.* Other feedstocks for ethanol production include milo, wheat, sorghum, barley, brewery waste products, and cheese whey. However, feedstock availability and policy preferences will limit their use over the time frame of this study.

Of the first-generation ethanol feedstocks, sugar cane and sugar beets have the highest yield in gallons per acre, 37 percent higher than corn. However, corn has fewer byproducts, and ethanol produced from U.S.-harvested sugar cane costs $2.77 per gallon compared with $1.46 per gallon from corn grain at current commodity prices.[69]

Corn

USDA reported that 86.5 million acres of corn were planted in 2009, yielding 12,086 million bushels (or 153.9 bushels per harvested acre).[70] The majority of this acreage is located in the Midwest.

Annual crop yields have increased over the last 10 years from 137.1 bushels per acre in 2000 to 162.9 in 2010.[71] USDA projects that advances in technology will further increase yields to 178.4 bushels per acre in 2020.[72]

USDA projects that the corn available annually for ethanol production will rise sharply from 3,677 to 5,025 million bushels from 2009 to 2020. This rise reflects projected corn production above a "baseline" demand for non-ethanol uses of corn of between 8.4 and 9.6 billion bushels from 2009 to 2020. From this growth in available corn—coupled with projected slight increases in the conversion rate of ethanol from corn—we project that the potential ethanol that can be produced from corn will increase 42 percent, from 10,606 to 15,075 million gallons between 2009 and 2020 (Figure 5-2 and Table 5-2).

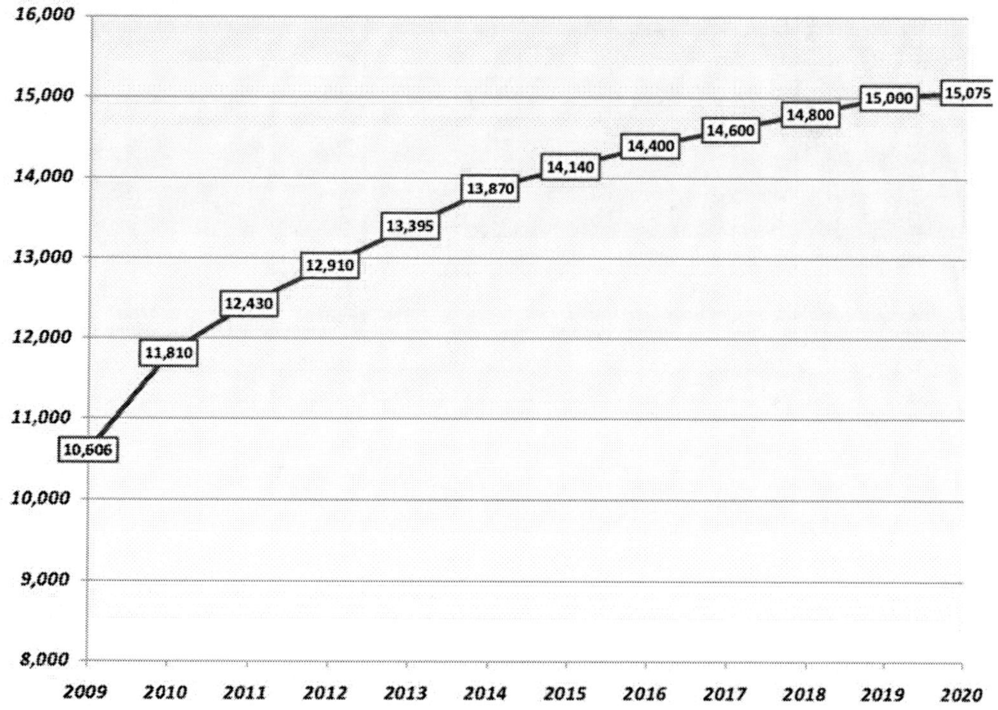

Figure 5-2. Forecast of Potential Ethanol Production from Corn, 2009–20 (million gallons).

Table 5-2. Forecast of Potential Ethanol Production from Corn, 2009–20

Measure	2009	2010	2011	2012	2013	2014	2015	2016	2017	2018	2019	2020
Million acres planted	86.5	86.4	88	90	89.5	89.5	89.5	89.5	89.5	89.5	89.5	89
Yield (bushels per acre)	153.9	162.9	160.4	162.4	164.4	166.4	168.4	170.4	172.4	174.4	176.4	178.4
Corn produced[a]	12,086	12,921	12,960	13,445	13,530	13,695	13,860	14,025	14,190	14,355	14,520	14,595
Corn available for ethanol use[a]	3,788	4,200	4,400	4,550	4,700	4,825	4,875	4,925	4,950	4,975	5,000	5,025
Corn available for other use[a]	8,424	8,721	8,560	8,895	8,830	8,870	8,985	9,100	9,240	9,380	9,520	9,570
Ethanol conversion rate[b]	2.8	2.81	2.83	2.84	2.85	2.88	2.9	2.93	2.95	2.98	3	3
Potential ethanol production[a]	10,606	11,810	12,430	12,910	13,395	13,870	14,140	14,400	14,600	14,800	15,000	15,075

[a] Millions of bushels.

[b] Gallons per bushel.

Less than half of the projected growth in corn production will be used to meet increased demand for ethanol production. Corn supply growth will result from both increased yields and acreage planted. We expect corn supply to exceed demand for ethanol use over the next 10 years, so corn prices will likely remain relatively stable over this period.

Prices

One of the key factors keeping ethanol cost-competitive with gasoline is the price of corn. According to DOE's *Annual Energy Outlook* (AEO), corn feedstock constitutes 57 percent of ethanol production costs.

The cost of corn depends on the supply and demand dynamics of the corn market. USDA forecasts that corn production will expand to meet ethanol requirements over the next 10 years, and corn prices will stabilize in 2013 near $3.70 per bushel.[73]

If corn prices behave as forecast, ethanol will continue to be cost-competitive with gasoline. Given current tax credits for ethanol production, Aventine Renewable Energy estimates that for ethanol to compete with gasoline at $30 per barrel of crude oil, corn prices must be less than $3.00 per bushel. At $40 per barrel of crude oil, ethanol is cost-competitive at corn prices less than $4.[74] If crude continues to cost well over $40 per barrel, a $4 per bushel cost of corn is probably sustainable.

Without tax credits for ethanol production, the landscape would change. Goldman Sachs estimates that if tax credits were taken away, new ethanol refinery builds would only be justified at crude oil prices above $63 per barrel (at projected longterm corn prices of $3.70 per bushel). If corn prices rise to $4 per bushel, ethanol production would expand only at crude oil prices above $65 per barrel. At $3.25 per bushel, such expansion would occur only at $55 per barrel and at $2.50 per bushel, expansion would occur at a crude oil cost above $45 per barrel.[75]

Availability for Food

Some argue that the increased percentage of corn used for ethanol raises a "foodversus-fuel" concern. However, the use of corn as a feedstock for ethanol has had relatively little effect on the availability of corn for food or the price of food. We estimate that due to increased corn harvest yields and ethanol conversion rates, the supply of corn for non-ethanol uses will increase by 14 percent from 2009 through 2020. USDA's June 2010 crop report suggests that increased production of ethanol is not having significant effects on the acreage of land dedicated to crops. Recent growth in the use of ethanol has only had limited impact on the retail price of food. The Congressional Budget Office (CBO) found that expanded production of ethanol between April 2007 and April 2008 resulted in an increase of 0.5 to 0.8 percentage points of the 5.1 percent increase in food prices measured by the consumer price index (CPI).[76] EPA, in the Final Rule on the Renewable Fuel Standard Program (March 2010), found that RFS2 will have an impact of $10 more per capita on food compared with the AEO 2007 reference case.[77]

In 2006, roughly 20 percent of the corn crop was used for ethanol. In 2010, due to rapid expansion in ethanol production, we estimate that approximately 32 percent of the crop will be used for ethanol. Our projections for corn-based ethanol production suggest that by 2020, this percentage will rise only by 2 percentage points, much less than previous increases.

Sugar Cane and Sugar Beets

Ethanol production from sugar cane and sugar beets is limited due to their high cost and limited availability relative to corn. Currently, no refineries in the United States are producing ethanol from sugars, and no new sugar refineries are planned.

In 2009, U.S. sugar cane production was 28.4 million tons and U.S. sugar beet production 29.5 million tons.[78] At this production level, total potential ethanol production using the entire harvest of domestic sugar cane and domestic sugar beets is only 554 million gallons and 732 million gallons, respectively. Combined, this theoretic production represents roughly 12 percent of the ethanol produced from corn in 2009.

Even if availability were not an issue, high feedstock and processing costs limit the current commercially viability of ethanol produced from sugar. At today's domestic sugar prices, the cost of converting sugar cane and sugar beets to ethanol is roughly $2.60 per gallon.[79] Feedstock costs represent between 65 and 70 percent of total production costs. With current (May 27, 2010) ethanol prices around $1.62 per gallon, producing ethanol from sugar is unprofitable.

Ethanol from sugar cane qualifies as an advanced biofuel for RFS2. However, to remain price competitive, feedstocks would have to be imported, and U.S. tariffs lowered. EPA estimates that 1.03 billion gallons of sugar cane ethanol will be imported to the United States in 2022 under RFS2.[80]

Second Generation

Second-generation biofuels include those produced from cellulose hemicellulose or lignin. Cellulosic feedstocks include corn stover, timber wastes, and dedicated energy crops such as switchgrass. Ethanol is produced from these feedstocks by isolating sugar molecules in the plant cell walls and converting them into ethanol. Technology enabling the production of ethanol from cellulosic materials is still in its early stages (several pilot plants are in operation), but the market is expected to grow dramatically over the next 10 years, driven primarily by RFS2.

Cellulosic biomass feedstocks—consisting of energy crops and agricultural, forest, and industrial residues—offer the greatest potential for producing the quantities of ethanol necessary to significantly displace petroleum. Some cellulosic crops have a higher yield per acre planted than corn—800 gallons compared with 416. Because these feedstocks can be grown on marginal lands with low energy, water, and fertilizer requirements, cellulosic biomass may be produced at much higher quantities and lower costs than corn.

Cellulosic Feedstock Availability

Estimates of the amount of biomass available for conversion into biofuel vary greatly. Table 5-3 shows the biomass estimates from three studies: (1) the DOE and USDA "billion-ton study,"[81] the EIA "25 × 25 study,"[82] and the Union of Concerned Scientists (UCS) "Climate 2030 study."[83] Table 5-4 shows the theoretical maximum potential biofuel that can be produced from the available cellulosic biomass.

Forest resources, urban wood waste, and mill residues available for cellulosic ethanol production are found in the wooded regions of the United States, including the Pacific Northwest, deep South, Canadian border regions in the Midwest, and Northeast.

Agricultural residues and energy crops are primarily located in the Corn Belt, Lake State, Appalachian, Delta, and Southeast regions. Corn stover production is dominant in the Corn

Belt and Lake State regions; straw in the Northern Plains, Mountain States, and Pacific regions; and energy crops in the Delta, Appalachian, Corn Belt, and Southeast regions.

Cellulosic Feedstock Costs

The economic competitiveness of cellulosic ethanol production is highly dependent on feedstock cost, which constitutes 35 to 50 percent of the total ethanol production cost, depending on v arious geographical factors and the types of systems used for harvesting, collecting, preprocessing, transporting, and handling the material.[84]

Table 5-3. Cellulosic Biomass Available for Conversion (million dry tons)

Biomass category	DOE/USDA billion-ton study	EIA 25 × 25 study	UCS Climate 2030 study
Agricultural residues	371–534	158	158
Energy crops	156–377	242	121
Forest residues	140	168	62
Urban wood waste and mill residues	63	29	27
Total available biomass	730–1,114	598	368

Table 5-4. Potential Biofuel Production from Available Cellulosic Biomass (billion gallons)

Biomass category	Conversion rate[a] (gal/ton)	DOE/ USDA billion-ton study	EIA 25 × 25 study	UCS Climate 2030 study
Agricultural residues	92.3	34.2–49.3	14.6	14.6
Energy crops	80	12.5–30.2	19.4	9.7
Forest residues	101.5	14.2	17.1	6.3
Urban wood waste and mill residues	90	5.7	2.6	2.4
Total potential production		66.6–99.4	53.7	33.0

[a] EPA, *Renewable Fuel Standard Program (RFS2) Regulatory Impact Analysis*, EPA-420-R-10-006, February 2010.

Table 5-5. Current and Projected Dry Herbaceous Cellulosic Feedstock Costs ($ per dry ton), 2009–17

Feedstock cost component	2009	2012	2017
Harvest and collection	12.40	12.15	10.80
Storage and queuing	6.75	5.95	5.30
Preprocessing	12.35	10.75	8.00
Transportation and handling	10.10	6.15	5.90
Total	41.60	35.00	30.00

Source: DOE, Office of the Biomass Program, Energy Efficiency and Renewable Energy, *Biomass Multi-Year Program Plan*, 2010, p. 3-21.

Table 5-6. Current and Projected Wet Herbaceous Cellulosic Feedstock Costs ($ per dry ton), 2009−17

Feedstock cost component	2009	2012	2017
Harvest and collection	20.70	10.60	10.60
Storage and queuing	17.80	11.10	8.60
Preprocessing	11.50	8.70	7.80
Transportation and handling	16.10	14.70	14.70
Total	66.10	45.10	41.70

Source: DOE, Office of the Biomass Program, Energy Efficiency and Renewable Energy, *Biomass Multi-Year Program Plan*, 2010, p. 3-21.

Cellulosic feedstocks are typically classified as wet or dry herbaceous, to address both storage degradation and transportation issues. Dry feedstocks, which have moisture contents less than 20 percent, include cereal straw and switchgrass. Wet feedstocks, which have moisture contents greater than 40 percent, include corn stover, sorghum stover, and forest residues.

Tables 5-5 and 5-6 provide estimates of current and projected feedstock costs for dry and wet feedstocks. The costs are broken out by each of the feedstock cost categories listed above.

Ethanol Production

First Generation

As of January 2010, 187 ethanol refineries (not including cellulosic ethanol refineries) were operating in the United States, representing a total production capacity of 11.9 bgpy (2009 production reached 10.6 billion gallons). Eleven ethanol refineries representing an additional production capacity of 1.2 bgpy are not currently operating. Ten new ethanol refineries are under construction and 5 are expanding, which will increase total production capacity to 14.4 bgpy over the next few years.[85] This increase in capacity will phase in over the next several years as construction projects are completed (Figure 5-3).

Currently, almost all ethanol refinery operating capacity uses corn as the primary feedstock—11,649 million gallons (or 98.1 percent) of the 11,877 million gallons of total capacity (including corn fraction when combined with other feedstocks).[86] This will continue in the foreseeable future because almost all of the idle capacity (98.7 percent) and most (89 percent) of the planned expansion of production capacity in the next few years will utilize corn feedstocks. As shown in Figure 5-3, after planned capacity comes online, ethanol production from corn (including corn fraction when combined with other feedstocks) will constitute 13,941 million gallons (or 97.2 percent) of the 14,346 million gallons of total refining capacity. Most of the current and planned refineries are located in the Corn Belt near the feedstock source.

Biorefineries produce corn-based ethanol using either dry or wet milling techniques (dry milling is the most common). Current dry milling techniques can produce 2.75 gallons of ethanol per bushel of corn (roughly 400 gallons/acre), while wet milling techniques yield 2.65 gallons per bushel (390 gallons/acre).

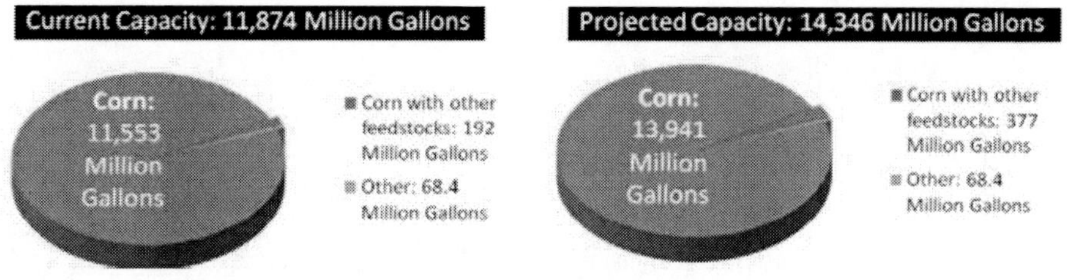

Figure 5-3. Current and Planned Capacity of Non-Cellulosic Ethanol Refineries by Feedstock.

Ethanol production from corn also generates byproducts that result in additional revenue for the refinery. In the dry milling process, roughly 6 pounds of dry distillers grains with solubles (DDGS), sold for use in animal feed, are produced per gallon of ethanol. The byproducts of the wet milling process include roughly 4.9 pounds of corn gluten feed, 0.9 pounds of corn gluten meal, and 0.6 pounds of corn oil per gallon of ethanol.[87]

Second Generation

Once feedstock is harvested, collected, stored, and preprocessed, it must be transported to a processing plant for conversion to fuel. The commercial cellulosic biofuel plants in the United States are limited today. As of January 2010, two cellulosic ethanol refineries were operating in the United States, representing a total production capacity of 3 million gallons per year (mgpy). One new cellulosic ethanol refinery is under construction with a production capacity of 100 mgpy.

We project growth in cellulosic ethanol plants in 2012 as technologies become commercially viable and cellulosic ethanol mandates increase. Until then, pilot plants will likely be in operation, as technologies to produce ethanol from cellulosic biomass are evaluated. In addition, smaller-scale "niche" cellulosic ethanol plants may be built to capture favorable economic opportunities. Figure 5-4 presents planned and existing locations and capacity for cellulosic biofuel (ethanol and biodiesel) plants.

Second-generation biofuels generally fall into two categories, which are based on the processes used to convert the biomass to fuel: biochemical or thermochemical. Biochemical conversion involves the fermentation of sugars from the breakdown of the biomass. Thermochemical involves the

> gasification and pyrolysis of biomass material into a synthesis gas or liquid oil for subsequent fermentation or catalysis. The main benefit of gasification and pyrolysis over the biochemical route is that thermochemical processes can more easily convert low-carbohydrate or "non-fermentable" biomass materials such as forest and wood residues to alcohol fuels and can more readily accept a wider variety of feedstocks. However, the thermochemical process does have some drawbacks, such as tar production and clean-up gas procedures that require additional capital investment.[88]

Commercial production of second-generation or cellulosic ethanol is still in its infancy; the predominant and most viable production process has yet to be determined. Planned production plants and ongoing research and development are supporting both production pathways.

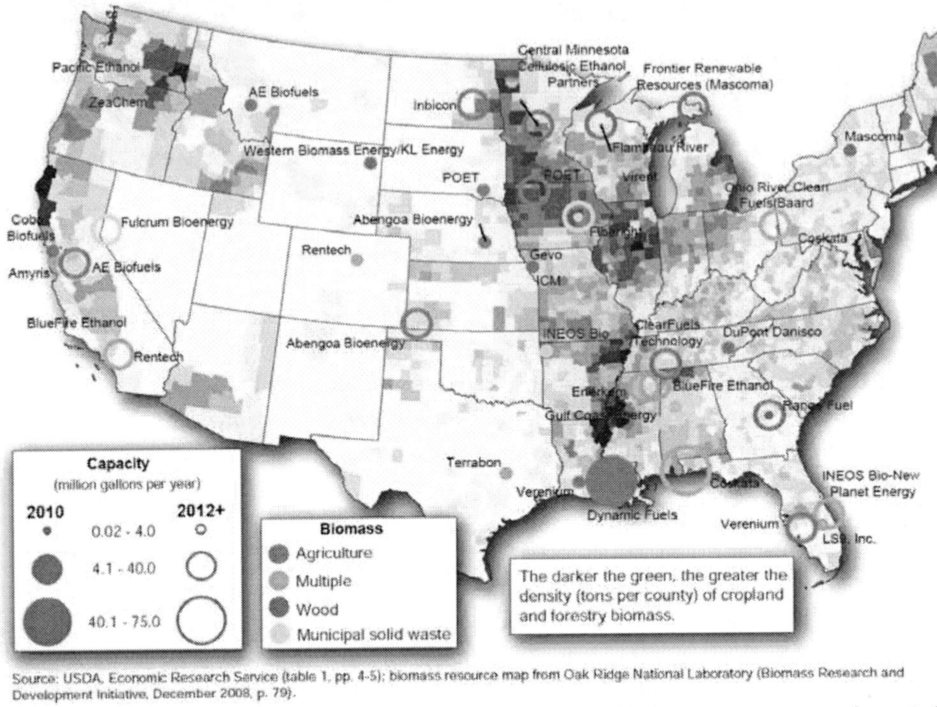

Source: USDA, Economic Research Service (table 1, pp. 4-5): biomass resource map from Oak Ridge National Laboratory (Biomass Research and Development Initiative, December 2008, p.79).
William T. Coyle, *Next-Generation Biofuels: Near-Term Challenges and Implications for Agriculture*, Economic Research Service, USDA, May 2010.

Figure 5-4. Locations of Existing and Planned Cellulosic Biofuel Plants.

Production Costs

First Generation

At the June 2010 price of corn ($3.24 per bushel), the net production cost per gallon of ethanol is $1.46.[89] Table 5-7 presents a breakdown of the production costs of ethanol.

Ethanol production costs may improve slightly due to recent technological improvements, including fractionation (oil extraction technologies), low heat fermentation, and use of pretreated or gray water.[90]

Table 5-7. Breakdown of Current Net Production Costs for Ethanol

Production cost component	Production cost ($/gallon)
Corn feedstock	1.16
Natural gas	0.18
Other variable costs	0.21
Fixed costs	0.21
Total production costs	1.77
Sales of byproducts (DDGS)	0.31
Total net production cost	1.46

Second Generation

Current costs for the production of cellulosic ethanol are more than $2 per gallon—Poet, LLC, current production cost for cellulosic ethanol produced fromcorn husks and cobs is $2.35 a gallon.[91] Production requires higher initial capital investment for biorefinery construction than for first-generation ethanol. As the technology develops and is refined, these investment requirements should decrease. Once plants are operating at economic scale and production processes have been refined, cellulosic ethanol production costs are projected to be lower than those of first-generation ethanol. Tables 5-8 and 5-9 show the breakdown of projected cellulosic ethanol costs for the biochemical and thermochemical processes. Both are less than the current net production costs of $1.46 per gallon for corn-based ethanol.

Table 5-8. Breakdown of Projected Biochemical Cellulosic Ethanol Production Costs (Corn Stover)

Production cost component	Production cost ($/gallon)	Percent
Biomass	0.51	38.5
Feed handling	Included above	
Pretreatment/conditioning	0.25	18.6
Hybrid hydrolysis and fermentation	0.10	7.8
Cellulase enzyme	0.10	7.5
Distillation and solids recovery	0.15	11.2
Wastewater treatment	0.02	1.8
Boiler/turbogenerator	0.13	9.8
Utilities	0.05	3.8
Storage	0.01	1.0
Total net production cost	1.33	100

Source: Thomas D. Foust, Andy Aden, Abhijit Dutta, and Steven Phillips, "An economic and environmental comparison of a biochemical and a thermochemical lignocellulosic ethanol conversion processes," *Cellulose* (2009), 16:547–565.

Table 5-9. Breakdown of Projected Thermochemical Cellulosic Ethanol Production Costs (Wood Chips)

Production cost component	Production cost ($/gallon)	Percent
Feedstock	0.57	46.9
Feed handling and drying	0.14	11.8
Gasification	0.11	9.1
Tar reforming: acid gas and sulfur removal	0.34	28.0
Alcohol synthesis—compression	0.10	7.9
Alcohol synthesis—other	–0.18	–14.6
Alcohol separation	0.05	3.7
Steam system and power generation	0.05	4.4
Cooling waters and other utilities	0.03	2.8
Total net production cost	1.22	100

Source: Thomas D. Foust, Andy Aden, Abhijit Dutta, and Steven Phillips, "An economic and environmental comparison of a biochemical and a thermochemical lignocellulosic ethanol conversion processes," *Cellulose* (2009), 16:547 565.

Tax Credit and Implications

In 2004, the American Jobs Creation Act (Public Law 108-357), established the Volumetric Ethanol Excise Tax Credit (VEETC) to provide a $0.51 per gallon payment for blending ethanol into the gasoline supply. The Food, Conservation, and Energy Act of 2008 (also referred to as the 2008 Farm Bill) reduced the ethanol tax incentive by 6 cents to $0.45 per gallon.

First Generation Ethanol Production Tax Credits

With the current tax credit of $0.45 per gallon, first-generation ethanol production costs are reduced to $1.01, and ethanol remains competitive with gasoline as a fuel additive. (The tax credit actually is provided directly to ethanol blenders, and it is implied in the comparison of production costs with gasoline.) As shown in Figure 5-5, the production cost of ethanol rises and falls with the price of corn. The current implied ethanol production price of $1.01 per gallon (with tax credit) is equivalent to gasoline production with crude oil at roughly $39 per barrel. Without the tax credit, the equivalent crude oil price jumps to $57 per barrel.

In addition to the VEETC, the Small Ethanol Producer Tax Credit (SEPTC, 26 *U.S. Code* 40) provides a tax credit for small ethanol production (annual capacity less than 60 million gallons) of $0.10 per gallon on the first 15 million gallons of ethanol produced in a tax year. The VEETC and SEPTC are set to expire on December 31, 2010.

Both the VEETC and SEPTC have been major factors in the increase in ethanol production from 3.4 to 10.6 billion gallons from 2004 to 2009. These tax credits provide "a safety net to ethanol producers and blenders against wide fluctuations in oil prices and thereby assures ethanol producers and investors that their product will remain cost competitive over sustained periods of time."[92] They also help support domestic ethanol production used to fulfill RFS2 requirements, rather than "enable foreign producers, such as Brazil, to export ethanol directly to the U.S. competitively."

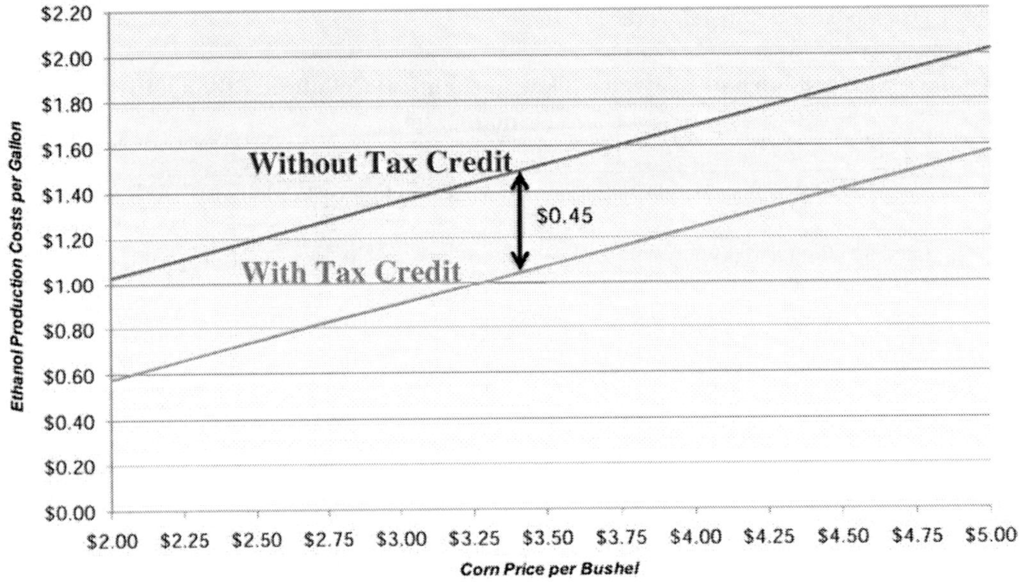

Figure 5-5. Net Ethanol Production Costs and Commodity Price of Corn.

Second Generation Cellulosic Ethanol Production Tax Credit

In May 2008, the Food, Conservation, and Energy Act of 2008 (Public Law 110234, Section 15321) established the Cellulosic Biofuel Producer Tax Credit (CBPTC). The CBPTC provides a tax credit of up to $1.01 for producers of cellulosic ethanol. (This credit includes the VEETC and the SEPTC credits as components.) The CBPTC—set to expire on December 31, 2012—is intended to support the early growth of the cellulosic ethanol industry.

Future Cellulosic Tax Credits

On March 25, 2010, Congress introduced legislation, the Grow Renewable Energy from Ethanol Naturally Jobs Act of 2010 (H.R. 4940, GREEN Jobs Act of 2010) extend VEETC, SEPTC, and CBPTC at their current levels through 2015. If these credits are not extended, there may be some short-term disruptions in the domestic ethanol industry as some higher cost producers close and a potential increase in ethanol imports.

Ethanol Transportation

There is an incongruity between where ethanol production takes place and feedstock supply with petroleum distribution infrastructure. Currently, one-third to one-half of ethanol travels through Chicago, which is a "choke point."[93] As biofuel usage increases, petroleum terminals will need to build accommodations, including upgrades to truck and barge receipt.

Denatured ethanol produced at ethanol refineries is transported to bulk terminals before final distribution to retail facilities. Ethanol is mostly transported by rail, barge, and tank truck from the Corn Belt (in the Midwest) to petroleum terminals (largely located on the coasts) for mixing and distribution.

Because it tends to separate from gasoline, ethanol is typically blended at distribution terminals, just prior to transportation to retail stations.[94] Therefore, storage requirements for ethanol prior to blending and transporting to retail stations can be significant. Because ethanol is more corrosive than gasoline, storage tanks must meet unique specifications.

The challenge to national distribution of ethanol is the limited capacity of the freight rail system, coupled with the lack of dedicated ethanol pipelines.[95] In the foreseeable future, a pipeline dedicated to ethanol transport is unlikely to be constructed, primarily due to high capital requirements relative to potential ethanol pipeline volume. About 60 percent of corn-based ethanol is transported by rail, 30 percent by truck, and 10 percent by barge.[96] Although the barge industry has sufficient capacity, this mode is limited due to lack of proximity of ethanol refineries to barge terminals and the limited scale of ethanol deliveries.

As of July 2010, there was a backlog of 14,930 freight cars, representing slightly less than 1 year of rail car production.[97] We estimate that 27,500 additional freight cars will be required to handle the projected increase in ethanol production by 2020,[98] or roughly 5 percent of all freight cars projected to be produced during this period.

Moving ethanol to the point of sale is far more expensive per gallon than gasoline. In June 2010, rail tank car transportation costs per gallon for ethanol (from refineries to fueling facilities) ranged from 8 to 26 cents, depending on distance and transportation mode, between two and five times the transportation cost for gasoline, an estimated 3 to 5 cents per gallon. Tank trucks are primarily used for delivery from the terminal to the retail infrastructure (short distance and lower volume requirements). Tanker truck is the most available but also the most costly ethanol transportation method: its average cost of 26.61 cents per ton-mile is almost 12 times that of rail tank cars.[99] Capacity, labor shortages, and public opposition to increased truck traffic are also issues.

Ethanol Retail Sales

The retail demand for ethanol and available production capacity drive its production. Most of the demand for ethanol in the United States is as an additive (1) to replace methyl tertiary butyl ether (MTBE), a suspected carcinogen implicated in the contamination of drinking water, and (2) as a relatively low-cost octane enhancer. As an additive, ethanol is blended with gasoline at 10 percent volume, referred to as E10. EPA is currently (FY10) evaluating whether to increase the maximum blend from E10 to E15.

Ethanol blended as E10 (or gasohol) currently accounts for 99.5 percent of all ethanol used in the United States and is expected to remain near this proportion through 2015. Almost all remaining ethanol production will be blended as E85. E85 retail sales are projected to increase dramatically in 2016, driven by the growth of cellulosic ethanol production coupled with limitations on the volume of ethanol that can be blended into gasohol. EPA estimates that in order to support our forecasts, consumers need to have reasonable access to E85 in 60 percent of the nation by 2022.[100]

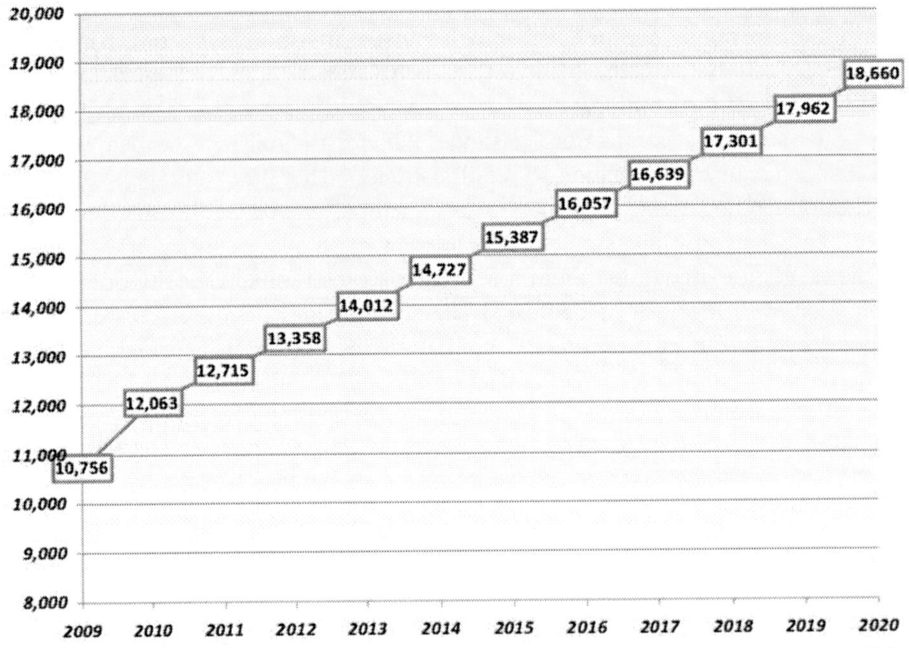

Sources: USDA *Agricultural Projections to 2019*, OCE-2010-1, February 2010; FAPRI at the University of Missouri, *U.S. and World Agricultural Outlook*, 2010; and EPA, *Renewable Fuel Standard Program (RFS2) Regulatory Impact Analysis*, EPA-420-R-10-006, February 2010.

Figure 5-6. Current and Projected Retail Sales of Ethanol, 2009–20 (million gallons).

Overall Forecast

We forecast retail ethanol sales using inputs from three forecast projections: USDA, EPA, and the Food and Agricultural Policy Research Institute (FAPRI). As shown in Figure 5-6, we project ethanol retail sales to increase more than 73 percent from 2009 to 2020, from 10,756 to 18,660 million gallons. Table 5-10 provides a breakdown of the ethanol retail sales by end-use fuel.

Gasohol (E10 or E15)

In 2009, E10 represented 79 percent of all gasoline sold in the United States and 75 percent or more of gasoline sales in 31 states. As production increases, ethanol retail sales will likely exceed the "E10 blending wall," or the theoretical maximum ethanol that can be blended into the projected demand for gasoline. Our forecasts show that ethanol production will exceed the E10 blending wall in 2013, (Figure 5-7). Before this occurs, we expect EPA to allow ethanol blends greater than 15 percent, either setting the maximum blend level at 12 percent or approving E15, but only for vehicles manufactured in 2001 and later. Therefore, the maximum E15 potential will likely never be realized, and the true market potential will be somewhere between 12 and 13 percent of the gasoline market.

Table 5-10. Forecast of Potential Ethanol Retail Sales, 2009–20 (million gallons)

	2009	2010	2011	2012	2013	2014	2015	2016	2017	2018	2019	2020
E10, E12, or E15	10,702	12,005	12,654	13,288	13,936	14,646	15,285	14,917	15,153	15,456	15,531	15,431
E85	54.0	57.6	61.3	69.3	75.8	80.5	101	1,139	1,486	1,844	2,431	3,229
Total	10,756	12,063	12,715	13,358	14,012	14,727	15,387	16,056	16,638	17,301	17,962	18,660

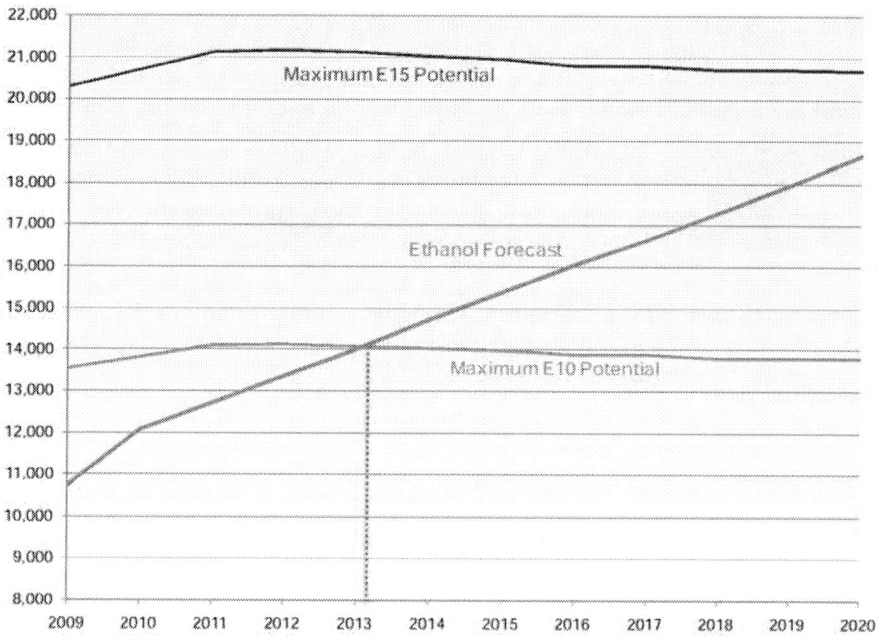

Figure 5-7. Market Potential for Ethanol as E10 and E15, 2009 20 (million gallons).

Table 5-11. Implied Wholesale Prices of E10 and E85 at Current Gasoline Prices

Category	Gasoline	E10	E85
Implied cost of ethanol at retail pump ($/gallon)			
Pump price	3.00	3.00	2.13
Taxes and margin	(0.60)	(0.60)	(0.60)
Implied fuel cost	2.40	2.40	1.53
Neat gasoline actual cost	2.40		
Sub-octane gasoline actual cost		2.16	0.36
Ethanol implied cost		0.24	1.17
Implied ethanol wholesale value			
Ethanol revenue per gallon of fuel sold at pump		$0.24	$1.17
Gallons of ethanol		0.10 gallon	0.85 gallon
Value of ethanol		$2.40	$1.38
Volumetric Ethanol Excise Tax Credit (VEETC)		$0.51	$0.51
Ethanol implied wholesale value/price		$2.91	$1.89

Source: Cliff Cook, Marathon Oil, *OPIS Ethanol and Biodiesel Supply Summit: Ethanol Expansion into Growth Markets*, March 2, 2007.

Economics also drive the demand for ethanol blended as E10—the market value of ethanol blended as E10 is far higher than that as E85. Because E85 has only 72 percent of the energy content per gallon of gasoline, ethanol blended as E85 must be discounted at the pump to account for its reduced fuel efficiency. However, because ethanol is a relatively cost-effective oxygenate or octane enhancer in gasoline, E10 is sold at the same price as gasoline. Therefore, at the current price of gasoline of $3.00 per gallon, the implied wholesale price of ethanol in E10 is $2.91 per gallon, or 54 percent higher than the implied wholesale price of ethanol in E85 of $1.89 per gallon. Table 5-11 shows the calculation of the implied wholesale prices for ethanol in E85 and E10.

The market for gasohol blends is far greater than E85 because gasohol is sold through the same pump and tank systems as gasoline. Therefore, gasohol can be made available at any of the 170,000 retail gasoline stations in the United States. As explained in the next subsection, E85 is limited to slightly more than 1 percent of those retail gasoline stations.

E85

DOE estimates that E85 is currently available at only 2,096 (or 1.2 percent) of the roughly 170,000 fuel stations in the United States.[101] Most of these stations are located near production facilities in the Midwest.

The availability of E85 is limited by the following:

- High infrastructure cost requirements
- Dispenser certification issues
- Low concentration of FFVs
- Price and fuel efficiency of E85 compared with gasoline
- Transportation issues.

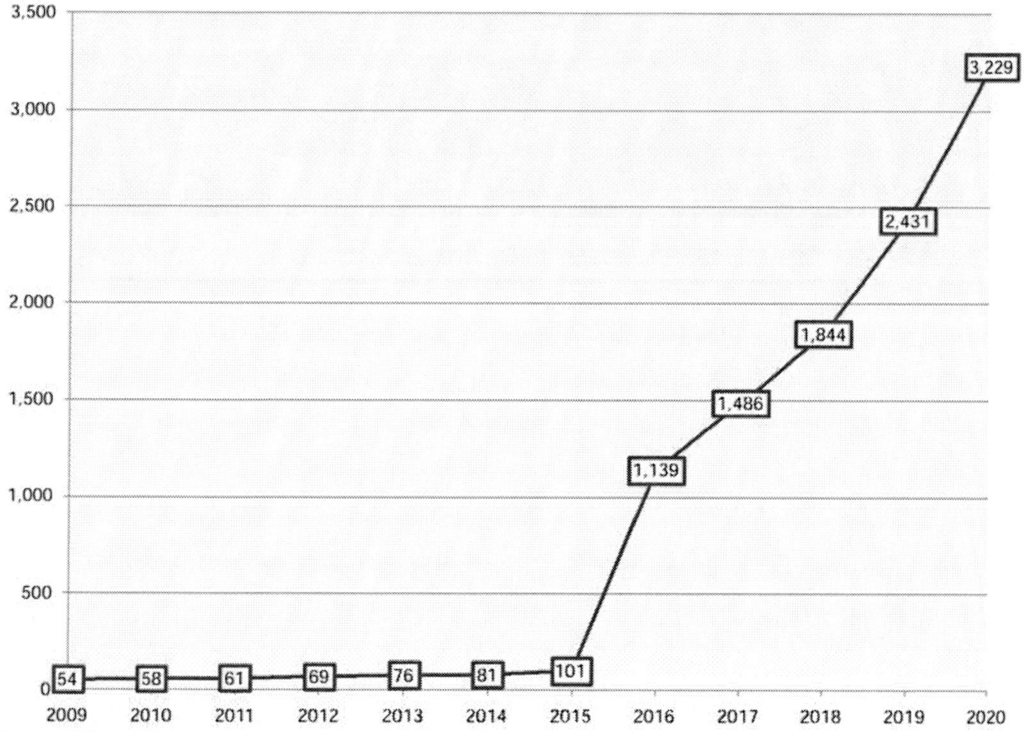

Source: EIA, AEO *2010*, DOE/EIA-0383(2010), May 11, 2010.

Figure 5-8. Current and Projected Retail Sales of E85, 2009–20 (million gallons).

We forecast retail sales of E85 on the basis of EIA data together with our forecasts of ethanol production and the E10 and E15 markets. As shown in Figure 5-8, we project that E85 retail sales will increase almost sixtyfold in 2009–20, from 54.0 to 3,229 million gallons.

Prices

Ethanol Commodity

The supply of and demand for ethanol production determine the price of ethanol. When demand and production are in equilibrium, the wholesale price of ethanol is tied to the wholesale price of unleaded gasoline, maintaining a $0.45 per gallon premium based on the tax credit.[102] However, when ethanol demand and supply are unbalanced, ethanol prices begin to deviate from gasoline prices. In cases of excess demand, ethanol prices increase relative to gasoline prices, and vice-versa in cases of excess supply.

As shown in Figure 5-9, increases in ethanol production capacity in 2006, 2007, and 2009 resulted in a temporary dip in ethanol wholesale prices below unleaded gasoline prices. As demand for ethanol in E10 began to catch up with production capacity, ethanol prices returned to equilibrium in 2010.

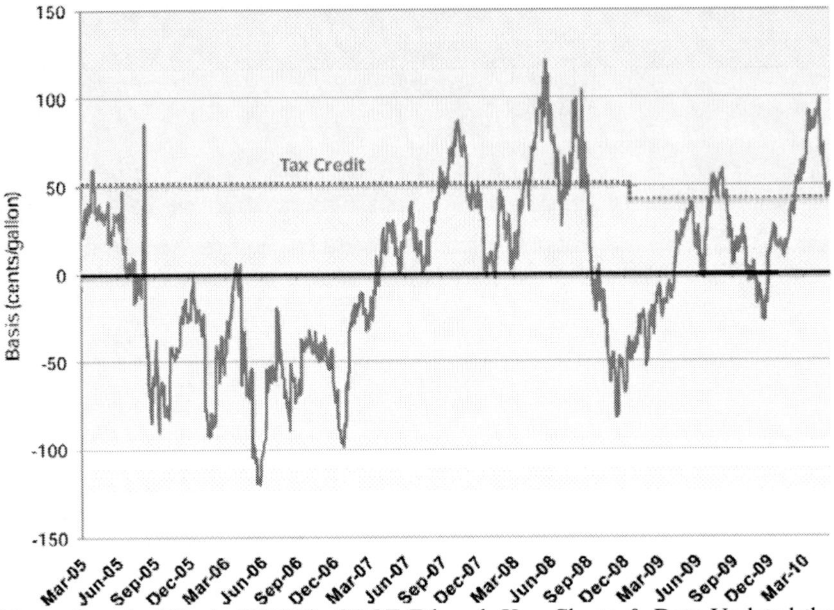

Source: Chicago Board of Trade (CBOT), CBOT Ethanol, Key Charts & Data Updated through June 2007, June 2007.

Figure 5-9. Spread between Ethanol and Unleaded Gasoline Prices.

Retail E85

Nationally, retail prices for E85 are lower than for regular unleaded gasoline, primarily to promote E85 use and account for E85's lower energy content (and fuel efficiency). In March 2007, DOE's Clean Cities Program reported that the national average price of E85 was $0.20 (or 9 percent) lower than regular gasoline.[103] With the exception of the Central Atlantic region, E85 was cheaper than gasoline, and the largest price differential ($0.29) was on the West Coast.

On the basis of its energy content, E85 should be priced at 72 percent of the regular gasoline price—it is currently priced at a premium to the consumer. Whether consumers will require a lower E85 price compared with gasoline in the future is unclear.

6. Biodiesel and Renewable and Cellulosic Diesel Market Projections

In this chapter, we present our assessment of the current and future commercial availability of biodiesel and renewable and cellulosic diesels. Today, most biodiesel is produced through FAME processes, which use vegetable oils or animal fats to produce biodiesel through transesterification. The current market for biodiesel has stagnated, primarily because of high retail prices compared with diesel fuel. However, we project the market for biodiesel will increase dramatically during the next 10 years, driven by the RFS2 mandate and the emergence of next generation renewable and cellulosic diesels.

To forecast the commercial availability of biodiesel and renewable and cellulosic diesels, we consider each supply chain stage. The supply chain includes feedstock production and distribution, fuel production, transportation of fuel to bulk facilities for blending with conventional fuels, and transportation to a retail facility for final sale to consumers. Market conditions, costs, and bottlenecks at each of these stages of the supply chain determine the availability of renewable fuels.

As shown in Table 6-1, we forecast that biodiesel and renewable and cellulosic diesel use will increase more than twelvefold to 5,848 million gallons by 2020. However, most of this growth will be driven by the renewable and cellulosic diesel markets. Throughout the next decade, the primary factor limiting FAME biodiesel availability will be retail demand. FAME biodiesel use will continue to be limited by high production costs and retail prices, its chemical properties in colder environments, inconsistent quality, and lack of consumer acceptance.

During the next 10 years, we forecast that a third-generation biofuel, FAME biodiesel produced from algal feedstocks, will emerge. However, high feedstock oil production costs will likely limit this biofuel to less than 100 million gallons through 2020. Third generation renewable fuels production systems, such as photosynthetic algae may have considerable water demand tradeoffs.

Within the next few years, we forecast that production of renewable and cellulosic diesel (typically produced from rendered animal fats and reclaimed cooking oils and greases) will increase. Biodiesel produced from these processes meets the fuel specification requirements of ASTM D975 (petroleum diesel fuel), and these drop-in fuels are anticipated to seamlessly blend into the diesel fuel stream. Because production costs and anticipated retail prices are competitive with conventional diesel, the primary factor limiting renewable and cellulosic diesel availability will be production capacity.

Biodiesel Supply Chain Overview

The supply chain for biodiesel is similar to that of ethanol (see Figure 6-1), except with different feedstocks. First-generation biodiesel includes current FAME biodiesel (a fuel comprising mono-alkyl esters of long chain fatty acids derived from vegetable oils or animal fats meeting the requirements of American Society for Testing and Materials (ASTM) D 6751 fuel) and renewable diesel (diesel fuel derived from biomass using a thermal depolymerization process that meets the requirements of ASTM D975 for petroleum diesel fuel). Renewable diesel is included in this category because of its anticipated early use of first-generation oil feedstocks (animal fats).

FAME biodiesel is produced from soybean oil, other vegetable oils, or animal fats through a process called transesterification, which separates glycerin (or glycerol) from oil. The process generates two primary products: glycerin and biodiesel (alkyl esters).[104] In the next few years, renewable diesel will increasingly emerge in the marketplace. Renewable diesel uses hydrotreating (or thermal depolymerization) processes to produce diesel fuel primarily from first-generation feedstocks, such as fats, waste oils, and greases, that compete for the same feedstock pool as current FAME biodiesel.

Table 6-1. Biodiesel and Renewable and Cellulosic Diesel Forecasts for Supply Chain Stages, 2009–20 (million gallons)

Stage	Type	2009[a]	2010	2011	2012	2013	2014	2015	2016	2017	2018	2019	2020
Feedstock availability	*FAME and Renewable Diesel*												
	Vegetable oils[b]	750	853	956	1,076	1,123	1,126	1,177	1,271	1,315	1,346	1,372	1,382
	Fats and greases[c]	1,205	1,205	1,205	1,205	1,205	1,205	1,205	1,205	1,205	1,205	1,205	1,205
	Total	1,955	2,058	2,161	2,281	2,328	2,331	2,382	2,476	2,520	2,551	2,577	2,587
	Second-Generation Biodiesel (Cellulosic Diesel)												
	Cellulose[d]	33.0 to 99.4 Billion Gallons											
Production capacity	*FAME Biodiesel*												
	Vegetable oils[b]	2,645	3,049	>3,049	>3,049	>3,049	>3,049	>3,049	>3,049	>3,049	>3,049	>3,049	>3,049
	Fats and greases[e,f]	45	51	70	88	107	131	144	167	189	211	233	256
	Algae[f]	0	0	0	0	0	0	0	10	20	40	60	80
	Total	2,690	3,100	>3,119	>3,137	>3,156	>3,180	>3,193	>3,226	>3,258	>3,300	>3,342	>3,385
	Renewable Biodiesel												
	Fats and greases[f]	0	0	44	89	89	126	167	167	167	167	167	167
	Cellulosic Biodiesel												
	Cellulose[f]	0	0	0	111	456	789	1,356	1,922	2,489	3,167	3,844	4,756
	Total	2,690	3,100	>3,163	>3,337	>3,700	>4,095	>4,716	>5,305	>5,893	>6,593	>7,293	>8,227
Projected production and retail sales	*FAME Biodiesel*												
	Vegetable oils[b]	434	598	724	846	914	914	956	1,037	1,081	1,107	1,120	1,108
	Fats and greases[b,f]	41	46	63	79	96	118	130	150	170	190	210	230
	Algae[f]	0	0	0	0	0	0	0	10	20	40	60	80
	Total	475	644	787	925	1,010	1,032	1,086	1,197	1,271	1,337	1,390	1,418
	Renewable Biodiesel												
	Fats and greases[f]	0	0	40	80	80	113	150	150	150	150	150	150

Stage	Type	2009[a]	2010	2011	2012	2013	2014	2015	2016	2017	2018	2019	2020
	Cellulosic Diesel												
	Cellulose[f]	0	0	0	100	410	710	1,220	1,730	2,240	2,850	3,460	4,280
	Total	475	644	827	1,105	1,500	1850	2,456	3,077	3,661	4,337	5,000	5,848

[a] 2009 values represent actual data.

[b] FAPRI, University of Missouri, *U.S. and World Agricultural Outlook*, 2010.

[c] National Biodiesel Board (NBB), *Biodiesel: Feedstock Supply, Achieving 1 Billion Gallons of Biodiesel While Protecting Valuable Feedstocks*, 2010.

[d] EIA, *Annual Energy Outlook 2010*, DOE/EIA-0383(2010), May 11, 2010.

[e] Biomass Research and Development Board, *The Economics of Biomass Feedstocks in the United States: A Review of the Literature*, Occasional Paper No. 1, October 2008.

[f] EPA, *Renewable Fuel Standard Program (RFS2) Regulatory Impact Analysis*, EPA-420-R-10-006, February 2010.

Second-generation biodiesel, commonly known as cellulosic diesel, is expected to emerge as renewable fuel within the next few years because of mandates and improved commercial viability. Cellulosic biodiesel is produced from cellulosic biomass using thermochemical processes, including the Fischer-Tropsch biomassto-liquids (BTL) process.[105]

Third-generation biodiesel includes FAME biodiesel (or potentially renewable diesel) fuels produced from advanced feedstocks, such as algal oil. Algal biodiesel offers many potential benefits, including minimal land requirements due to high oil yields, producibility using marginal land, and low-quality water inputs (brine water). However, large quantities of water are needed to grow the algal feedstock. These fuels are in their infancy and may not be commercially viable before 2020.

In 2008, approximately 720 million gallons of biodiesel were produced in the United States, almost all from vegetable oils.[106] In 2009, after the worldwide economic recession, production was at 475 million gallons.[107] Figure 6-1 shows the rapid growth of production between 2001 and 2008 and the decline in 2009.

As of June 2009, 173 biodiesel production companies were operating in the United States with an annual capacity of 2.69 billion gallons.[108] Approximately 427.8 million gallons of capacity is currently under construction, including the expansion of an existing plant.[109] The 29 companies constructing new biodiesel plants project completion by the end of 2010.[110] The estimated annual capacity to produce biodiesel will surpass 3.1 billion gallons by the end of 2010.

Like ethanol, FAME biodiesel cannot be transported through existing multiproduct pipelines. Because of this, according to EIA, "railroad cars and tanker trucks made from biofuel-compatible materials are needed to transport large volumes of biofuels to market."[111] FAME biodiesel is typically blended as B2 (2 percent biodiesel and 98 percent diesel), B5 (5 percent biodiesel), B20 (20 percent biodiesel), or B100 (pure biodiesel).[112]

However, both renewable and cellulosic diesels have a similar chemical composition to regular diesel and meet the diesel fuel specification requirements. As such, they can be transported through the same supply chain as conventional diesel. Therefore, these fuels do not raise the same transportation and other supply chain issues as FAME biodiesel. Renewable and cellulosic diesel can be transported through the same pipelines, use the same refueling infrastructure, and be used in the same vehicles as conventional petroleum diesel fuel.

Retail sales of FAME biodiesel are projected to increase roughly threefold by 2020. However, biodiesel only represents a very small fraction (less than 1 percent in 2007) of U.S. diesel transportation demand.[113] USDA and FAPRI predict FAME biodiesel production in the United States will increase to more than 1 billion gallons by 2012 or 2013.[114,115] Although not a significant portion of diesel demand, FAME biodiesel is nonetheless a growing component of the biofuels picture. The future growth of FAME biodiesel depends on its acceptance by consumers and market demand for diesel. If biodiesel becomes more accepted by diesel fleet operators, production may increase.

We forecast that biodiesel retail sales growth will be driven by the emergence of renewable and cellulosic diesel. By 2015, renewable and cellulosic diesel production volumes will exceed FAME biodiesel. By 2020, these fuels will account for almost 76 percent of biodiesel retail sales.

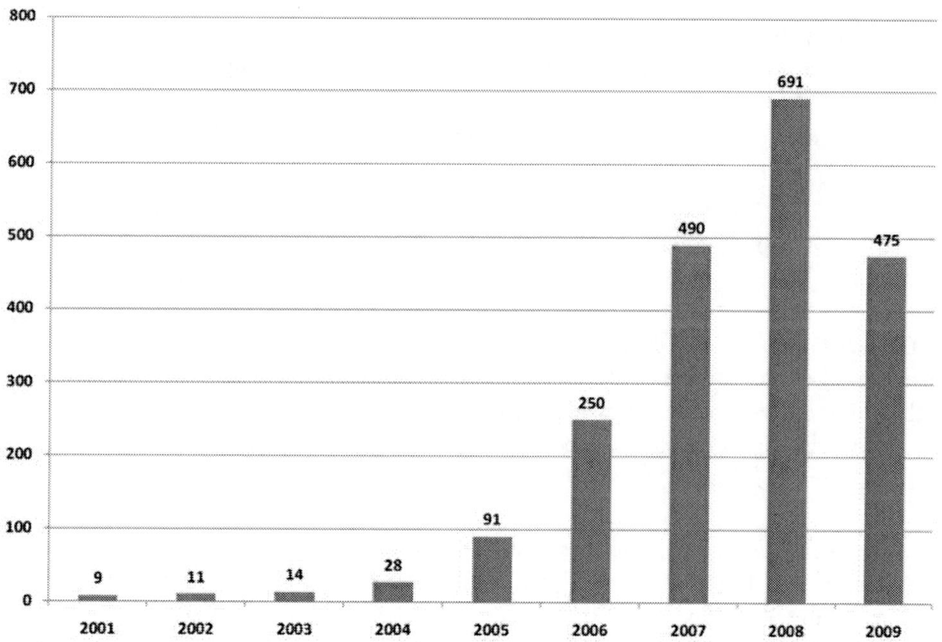

Sources: EIA, *Biodiesel Supply and Consumption, Supplement to the Short-Term Energy Outlook*, April 2009, www.eia.doe.gov/emeu/steo/pub/special/2009_sp_01.pdf; FAPRI, *U.S. and World Agricultural Outlook*, January 2010, FAPRI Staff Report 10-FSR 1, ISSN 1534-4533, Iowa State University, University of Missouri-Columbia, Ames, IA (FAPRI6US).

Note: Fuel quantities represent "neat," rather than blended, fuel.

Figure 6-1. Biodiesel Production 2001–09 (million gallons).

Biodiesel Feedstocks

Our forecast for feedstocks for biodiesel production through 2020 is as follows:

- *Vegetable oils will continue as the feedstock for roughly 85 percent of first-generation biodiesel.* Soybeans and other vegetable oils will remain the preferred feedstocks for first-generation biodiesel production because of their availability, commodity cost, and economics for production.
- *Biodiesel produced from fats, waste oils, and greases will grow almost tenfold by 2020.* Production costs for biodiesel using fats, waste oils, and greases are often lower than for vegetable oil feedstocks. "Much of this biodiesel production, however, seems to rely on niches of feedstock availability and market outlets."[116] EPA projects that by 2020, 230 million gallons per year of rendered or reclaimed fats will be processed by firstgeneration biodiesel plants and 150 mgpy will be used by renewable diesel facilities.[117]
- *The cellulosic pathway will be commercially viable after 2012 and grow rapidly through 2020.* Driven by RFS2, diesel produced from cellulosic feedstocks will begin in 2012 and grow to more than 5 billion gallons by 2020. Dedicated energy crops (such as switchgrass, energy cane, and biomass sorghum) are forecast to

constitute roughly 65 percent of the cellulosic feedstocks, with 21 percent crop residues (corn stover and straw) and 14 percent woody biomass (primarily logging residues).

- *Algal FAME biodiesel will emerge in 2016, but will be limited in commercially viability until after 2022.* EPA projects limited commercially viable FAME biodiesel production (100 million gallons) from algal feedstocks by 2022, but it does not forecast any production of ethanol or diesel fuel from algal feedstocks before 2022.[118]

First-Generation Biodiesel Feedstocks

Almost all of the biodiesel produced today is considered a first-generation biofuel, or a biofuel generated from crops or animal fats that can also be used as food and feed. The primary potential first-generation feedstocks for biodiesel production in the United States through 2020 are as follows:

- *Soybean oils.* Until 2006, soybean oil represented the feedstock for almost all (97 percent) of biodiesel production in the United States. Between FY07 and FY09, soybean oil dropped to 45 percent of biodiesel production, primarily because of a spike in soybean prices during this period. We forecast that soybean oil feedstock will constitute 48 to 53 percent of first-generation biodiesel production through 2020.
- *Other vegetable oils.* Corn oil (including oil extracted during ethanol production) and canola oil are the primary other vegetable oils used for biodiesel production. In 2009, other vegetable oils represented roughly 39 percent of biodiesel production. Other vegetable oil feedstocks are forecast to decrease from 46 percent of first-generation biodiesel production in 2010 to 30 percent in 2020.
- *Animal fats and restaurant grease.* Rendered animal fats and reclaimed cooking oils and greases represent the other major source of biodiesel feedstock. In 2009, animal fats and restaurant grease represented roughly 9 percent of biodiesel production. We forecast that these feedstocks will increase to constitute more than 17 percent of first-generation biodiesel production through 2020.

First-Generation Biodiesel Feedstock Yields and Costs

The economics of biodiesel production primarily depend on the cost of the feedstock. Feedstock costs are the largest component of biodiesel production costs: the Iowa State Center for Industrial Research and Service estimates that feedstock costs constitute 72 percent of total biodiesel production costs.[119] Table 6-2 and Figure 6-2 compare wholesale production costs for first-generation biodiesel produced from different feedstocks.

Within the vegetable oil feedstocks (soybean, corn, canola, cottonseed, peanut, and sunflower oils), soybean and corn are the most cost competitive for biodiesel production. Although production costs from fats are lower than those for vegetable oils, supply availability and potential fuel quality issues limit the potential from these feedstocks.

With the exception of yellow grease, feedstock costs per gallon of biodiesel are significantly higher than feedstock oil cost per gallon of diesel. Currently, only high subsidy levels enable biodiesel to remain cost competitive with diesel in retail markets.

Table 6-2. Comparison of Biodiesel Production Costs by Feedstock

Feedstock	Wholesale feedstock price ($ per pound)	Wholesale feedstock price ($ per gallon of biodiesel)
Soybean oil	0.3683	2.76
Corn oil	0.3696	2.77
Canola oil	0.4461	3.35
Cottonseed oil	0.4212	3.16
Peanut oil	0.6207	4.66
Sunflower oil	0.5565	4.17
Tallow	0.2775	2.08
Yellow grease	0.2400	1.83

Sources: FAPRI, *U.S. and World Agricultural Outlook*, January 2010, FAPRI Staff Report 10-FSR 1, ISSN 1534-4533, Iowa State University, University of MissouriColumbia, Ames, Iowa (FAPRI6US); Fred Wellons, National Renderers Association, *Transforming Animal Fats and Used Cooking Oils into Green Fuels—Technology that Works*, March 16, 2010.

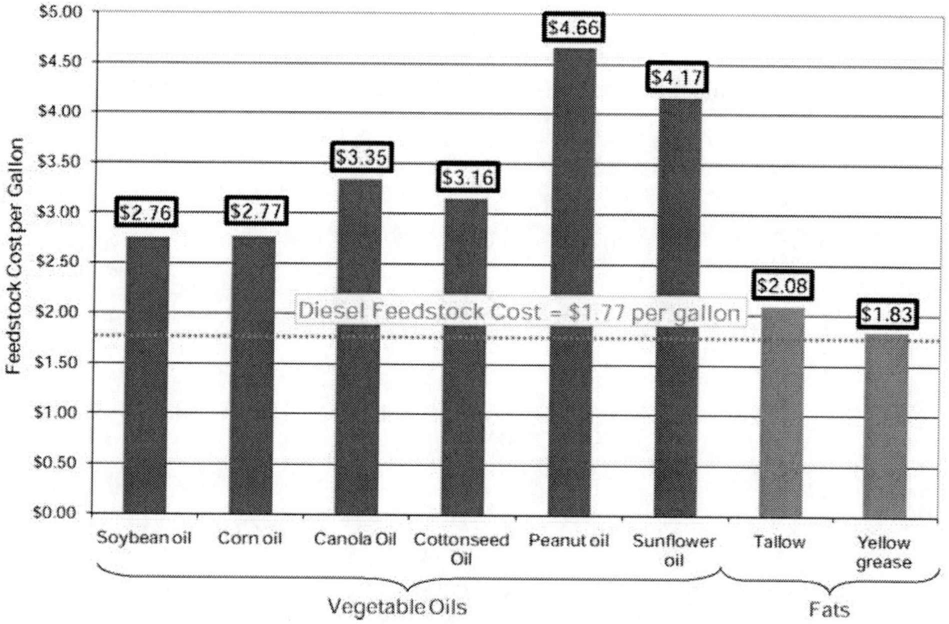

Figure 6-2. Comparison of Biodiesel Production Costs by Feedstock.

Soybean Oil

USDA reported that in 2009, 75.7 million acres of soybeans were planted, yielding 2,967 million bushels.[120] The majority of this acreage is located in the Midwest, the Mid-Atlantic states, and along the Mississippi River.

Annual crop yields have increased during the last 10 years, from 38.1 bushels per acre in 2000 to a projected 44 in 2010.[121] USDA projects that advances in technology will further increase yields to 46.5 bushels per acre by 2020.[122]

FAPRI projects that soybean oil available annually for biodiesel production will rise sharply from 1,904 million pounds in 2009 to 5,412 million pounds in 2020. This rise reflects projected soybean oil production above a "baseline" demand for non-biodiesel uses of soybean oil between 14.2 billion pounds in 2009 and 15.7 billion pounds in 2020. From this growth in available soybean oil, we project that the potential biodiesel that could be produced from soybean oil will increase almost threefold, from 254,000 gallons in 2009 to 722,000 gallons in 2020 (see Figure 6-3 and Table 6-3).

The EIA has the following explanation why soybean oil dominates as a feedstock for biodiesel production:

> Soy is a v ersatile, nitrogen-fixing crop that yields oil and food for humans and livestock. Soybean meal is of higher market value than soy oil. Consequently, soy oil is a low-priced byproduct available in relatively large volumes. Currently, it is a cheaper virgin feedstock than other oilseeds. The processing and distribution infrastructure for soybeans is already in place, with more capacity being added as more biodiesel production facilities come online.[123]

Production of biodiesel from soybeans is not expected to significantly impact the food markets in the near term. However, if the primary feedstock for biodiesel production again becomes soybean oil, and demand for biodiesel continues to grow, an increased demand for biodiesel soybean oil could increase the wholesale price of soybeans toward the end of the decade.

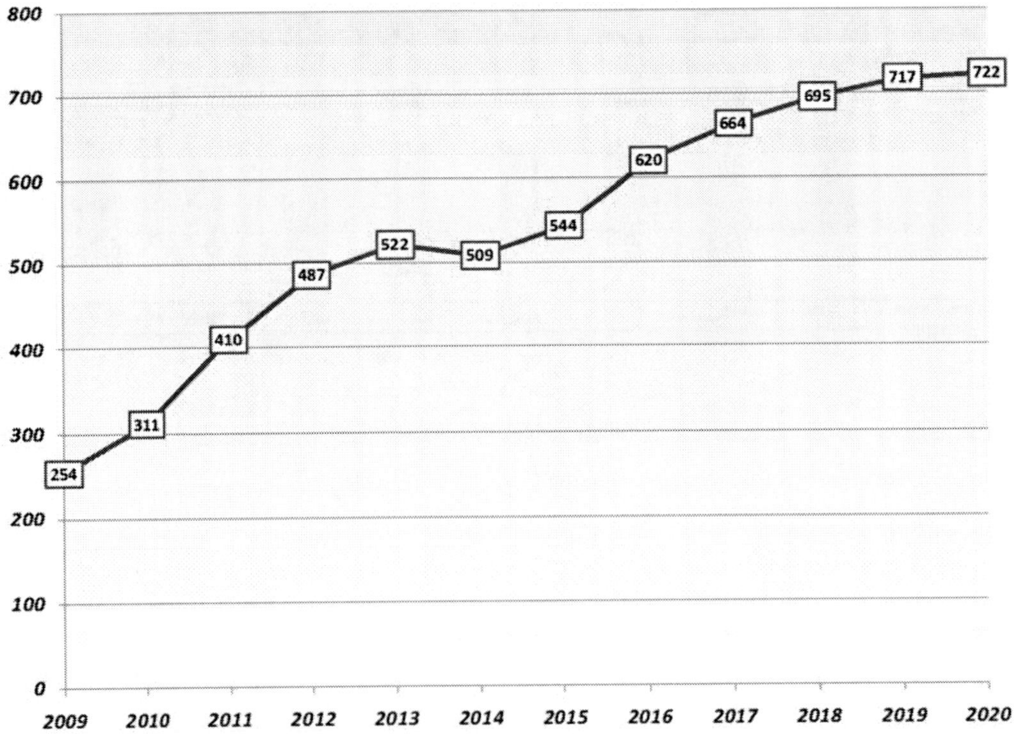

Figure 6-3. Forecast of Potential Biodiesel Production from Soybeans, 2009–20 (thousand gallons).

Table 6-3. Forecast of Potential Biodiesel Production from Soybeans, 2009–20

Category	2009	2010	2011	2012	2013	2014	2015	2016	2017	2018	2019	2020
Million acres planted	75.7	77.5	76.5	73.5	74.5	75.5	76.0	76.0	76.0	76.0	76.0	76.0
Yield (bushels per acre)	39.7	43.3	42.8	43.2	43.6	44.4	44.4	44.9	45.3	45.7	46.4	46.5
Soybeans produced[a]	2,967	3,319	3,230	3,130	3,205	3,280	3,330	3,370	3,400	3,430	3,460	3,490
Soybean oil production[b]	18,746	19,072	20,301	20,714	21,150	21,490	21,827	22,212	22,612	22,988	23,358	23,742
Soybean oil for food use[b]	14,481	14,222	14,562	14,459	14,608	14,792	14,934	15,041	15,171	15,309	15,474	15,666
Soybean oil for biodiesel use[b]	1,904	2,335	3,077	3,655	3,911	3,817	4,080	4,650	4,982	5,210	5,380	5,412
Potential biodiesel production[c]	254	311	410	487	522	509	544	620	664	695	717	722

[a] Millions of bushels.
[b] Millions of pounds.
[c] Thousands of gallons.

Table 6-4. Forecast of Potential Biodiesel Production from Other Vegetable Oils, 2009–20 (million gallons)

	2009	2010	2011	2012	2013	2014	2015	2016	2017	2018	2019	2020
Canola oil	44	46	41	49	48	48	50	53	51	49	48	49
Corn oil	352	392	412	428	444	460	469	478	484	491	498	500
Other oils	100	105	93	111	109	109	114	120	116	111	109	111
Total potential other vegetable oils	496	542	546	589	601	617	633	651	651	651	655	660

Increased biodiesel production has already affected the market for glycerin. Roughly 10 pounds of crude glycerin is generated as a coproduct for every 100 pounds of biodiesel production. The amount of glycerin generated by the biodiesel industry has resulted in substantial oversupply (amounting to most of the glycerin produced in the United States) and a drop in the price of crude glycerin.[124]

Other Vegetable Oils

The primary vegetable oils other than soybean oil used for biodiesel production include canola oil and oil extracted from corn or its fermentation coproducts in the dry mill ethanol production process. Table 6-4 shows our forecast of the potential biodiesel production from other vegetable oils.

Canola Oil

Canola is a type of rapeseed that contains roughly 40 percent oil. USDA reports that in 2009, 1.01 million acres of canola seed were planted, yielding 1,058 million pounds of canola oil. Because U.S. demand for canola exceeds domestic production, more than 61 percent of the canola oil used in the United States is imported, primarily from Canada. Canola is grown during the spring and winter, with winter yields significantly higher than those for spring (1,500 pounds per acre compared with 3,500 pounds per acre).[125]

Most canola is manufactured as an edible oil due to its low saturated fat and high omega-3 fatty acid content. This market, combined with a higher feedstock price compared with other alternatives, limits canola oil's current use in biodiesel production (roughly 44 million gallons in 2009). However, its future use as a biodiesel feedstock may increase because of its potential for growth as a winter crop and its high oil yields.

Corn Oil

Large increases in corn-based ethanol production may support the growth of corn oil generated from ethanol production as a biodiesel feedstock. For existing wet mill corn ethanol plants, corn oil is already generated from the production process. For dry mill corn ethanol plants, "extraction of oil from the thin stillage or distillers' grains with solubles (DGS) streams is a proven technology that can be retrofitted into existing plants relatively cheaply."[126] EPA estimates that approximately 70 percent of all ethanol plants will implement corn oil extraction systems by 2022, generating up to 680 million gallons per year of corn oil.[127]

Fats and Greases

Rendered animal fats and reclaimed cooking oils and greases provide a large potential feedstock for biodiesel production through 2020. The National Biodiesel Board (NBB) estimates that the annual volume of animal fats and restaurant greases available for biodiesel production are 905 million gallons and 300 million gallons, respectively.

Second-Generation (Cellulosic) Diesel Feedstocks

Second-generation biodiesel includes cellulosic diesel produced from cellulose hemicellulose or lignin. Cellulosic feedstocks include corn stover, timber wastes, and dedicated energy crops such as switchgrass. Technology supporting the production of diesel from cellulosic materials is still in its early stages, but the market is expected to grow dramatically during the next 10 years. That growth will be driven primary through RFS2. We forecast that cellulosic feedstocks will support the production of between 33.0 and 99.4 billion gallons of cellulosic biofuels by 2020. (See Chapter 5 for a discussion of the availability of cellulosic feedstocks.)

Third-Generation (Algal) Biodiesel Feedstocks

Third-generation biodiesel includes all biodiesel or diesel fuel produced from algal feedstocks. Algae used for biofuel production include diatoms, green algae, golden-brown algae, prymnesiophytes, eustigmatophytes, and cyanobacteria—all single-celled algae species that grow rapidly to form biomass-containing oils and residuals used for fuel production.[128] Algae offer an advantage as a biofuel feeds-tock because its oil yield per acre can be significantly higher than that of other potential biofuel feedstocks (Table 6-5).

As Figure 6-4 shows, algal oils can be used to produce a variety of biofuels, including FAME biodiesel, renewable diesel, ethanol, and biobutanol, depending on conversion process. EPA forecasts that only the FAME biodiesel production pathway will be commercially viable before 2020.

The main external factors affecting algae-based oil production are climate, water and nutrients source, and CO_2 source. For photosynthetic pathways, most algae strains require sufficient access to sunlight during the producing period (typically year-round) in order to be productive. Other climate factors, such as temperature, precipitation, and evaporation, also impact the productivity of algae when using pool and external bioreactor production approaches (less of a factor with dark reaction bioreactors). One of the benefits of algae is that wastewater, saline, or brackish water can be used to effectively grow algal cultures because the algae can absorb the nutrients contained in these types of waters. However, these production systems require freshwater inputs to prevent harmful levels of salt from developing and can have wastewater hazards to mitigate. Algae also require a CO_2-rich environment to achieve optimal growth.

Although the algal-based biofuels can be produced in the United States, some challenges exist. For example, although the United States has regions where optimal algae growth can be achieved, these areas do not, for the most part, have easy access to environmentally beneficial sources of water and CO_2.

Table 6-5. Biofuel Yields from Various Feedstocks

Crop	Oil yield (gal/acre/year)[a]
Corn	18
Soybean	48
Canola	127
Jatropha	202
Coconut	287
Oil palm	635
Algae	1,000–4,000

Sources: DOE Biomass Program, *National Algal Biofuels Technology Roadmap*, 2010; Yusuf Chisti, "Biodiesel from Microalgae," *Biotechnology Advances*, February 2007.

[a] Converted from liters/hectare.

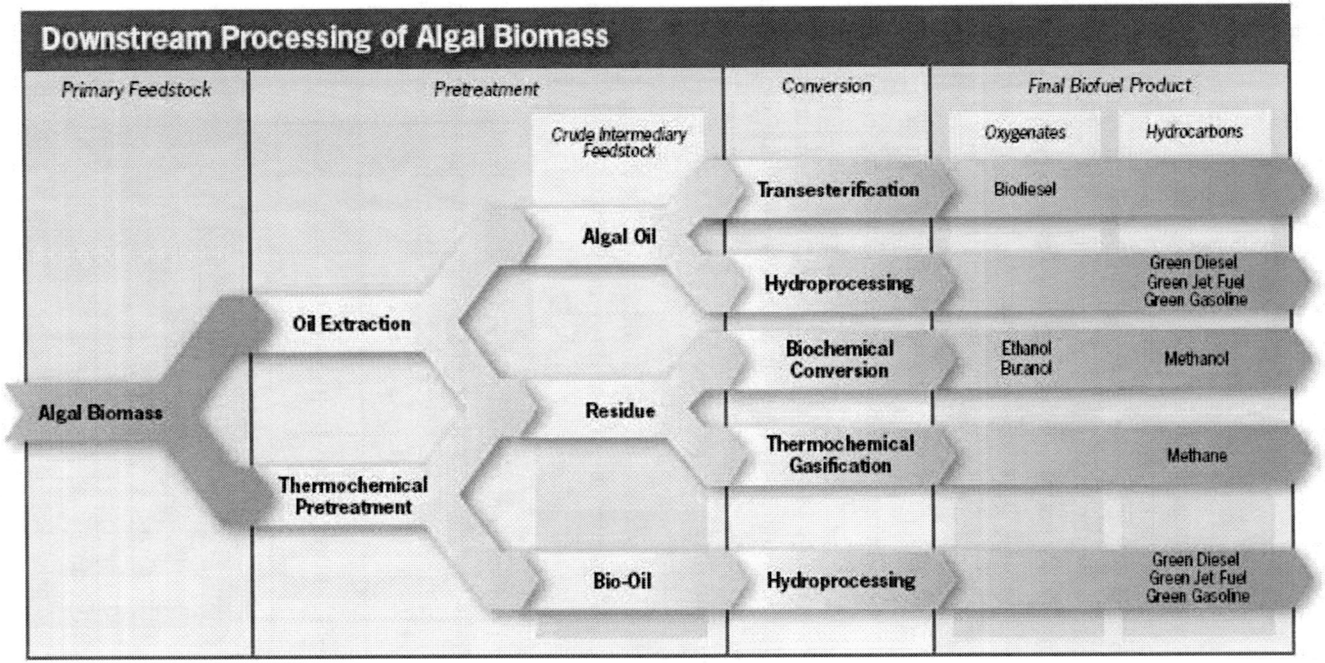

Source: Catie Ryan, Natural Resources Defense Council and Terrapin Bright Green, LLC, *Cultivating Clean Energy: The Promise of Algae Biofuels*, October 2009.

Figure 6-4. Pathways for Downstream Processing of Algal Biomass.

Biodiesel Production

First-Generation Biodiesel Production

Biodiesel annual production capacity currently exceeds the amount of biodiesel produced.[129] The National Biodiesel Board reports that the annual operational capacity of the biodiesel industry in the United States was 2.69 billion gallons (173 companies) as of June 2009. Approximately 427.8 million gallons of new capacity is under construction; 29 companies project completion of new construction by the end of 2010.[130]

As shown in Figure 6-5, less than half of all biodiesel refinery capacity is currently tied to soybean oil as the primary feedstock—approximately 1.10 billion gallons (41 percent) of the 2.69 billion gallons of total capacity. The majority of capacity, multi-feedstock refineries with 1.40 billion gallons (52 percent), also likely uses soybean oil as a primary feedstock, though they are able to process vegetable oils, animal fats, and restaurant grease. Less than 200 million gallons of capacity is devoted to canola oil and other oils and fats.[131]

Most biodiesel refineries are located in the Midwest, Southeast, and Mid-Atlantic regions, near their primary feedstock sources, but they are also concentrated in areas such as California and Texas.

FAME Process

Most first-generation biodiesel is currently produced using the FAME process, in which plant oils and animal fats (triglycerides) are chemically reacted with an alcohol (primarily methanol). The production process is very efficient (98 percent conversion) and creates glycerin (used in pharmaceuticals and cosmetics) as a by-product.

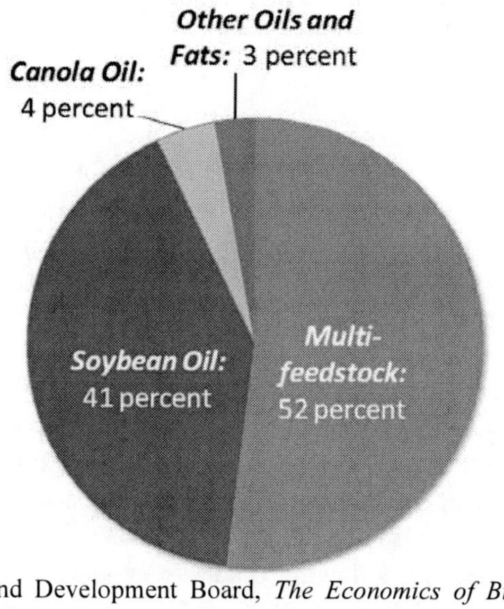

Source: Biomass Research and Development Board, *The Economics of Biomass Feedstocks in the United States: A Review of the Literature*, Occasional Paper No. 1, October 2008.

Figure 6-5. Capacity of Current Biodiesel Production Plants by Feedstock.

Renewable Diesel

Renewable diesel is an emerging biodiesel production process that converts vegetable oils and animal fats into diesel fuel using thermal depolymerization (a process currently used in petroleum refining). Although additional processes can produce a variety of petroleum products, EPA assumes roughly 90 percent of yield to diesel, with the remainder divided between light fuel gas and naphtha.[132]

Moreover, Dynamic Fuels, LLC, is expected to begin production of renewable diesel in Geismar, LA, within the next few years. This 75-million-gallon-per-year plant will use Tyson meat processing fats as feedstock to produce diesel fuel using hydrogen supply available locally.

HRJs are produced concurrently along with renewable diesels. While Chapter 8 discusses some of the additional early adopter biomass-based feedstocks, HRJs are generally a coproduct of renewable diesel feedstocks and production.

Second-Generation (Cellulosic) Biodiesel Production

No commercial cellulosic diesel plants are operating today—most existing plants are pilot or demonstration plants, producing fewer than 1 million gallons annually. However, we forecast growth in cellulosic diesel plants in 2012 when technologies become commercially viable and cellulosic fuel mandates increase. Until then, pilot plants will likely continue in operation while technologies for producing diesel from cellulosic biomass are evaluated. In addition, smaller-scale "niche" cellulosic diesel plants may be built to capture favorable economic opportunities.

Cellulosic diesel will likely be produced via two major thermochemical pathways: BTL (Fischer-Tropsch) and pyrolysis. These pathways generate a "drop-in" diesel fuel (pure hydrocarbon chains without ester links) that can be easily integrated into existing infrastructure.

Third-Generation (Algal) Biodiesel Production

The production of biofuels from algal feedstocks has four main steps: (1) a growth stage, where the algae to be harvested are allowed to grow in a medium (either fresh or impaired water); (2) a harvest stage, where the algae is collected; (3) an extraction stage, where the harvested algae is processed into biofuel feedstock; and (4) a production stage, where the feedstock is converted into biofuels (typically biodiesel or synthetic crude oil for refining into renewable diesel or jet fuel).

The current market for algal-based biofuels is in its infancy. Most current production in the United States is relegated to small pilot plants and to university and industrial research efforts. One major research initiative started by industry in 2009 is not expected to begin yielding results until 2014 at the earliest.[133] This situation makes it difficult to forecast the growth of this market.

In 2009, DOE awarded the following grants to support commercialization of biofuels from algal feedstocks: Algenol Biofuels ($25 million for a pilot project in Freeport, TX); Solazyme ($22 million for a pilot project in Riverside, PA); and Sapphire Energy ($50 million for a demonstration project in Columbus, NM).[134]

Many companies have announced plans for commercial-scale biofuel production efforts using algal feedstocks, but projections using these announcements are highly uncertain. Planned algae biofuel production before 2020 includes Sapphire Energy (synthetic crude oil to diesel and jet fuel), Petrosun (FAME biodiesel), Solazyme (FAME biodiesel and synthetic crude oil to diesel and jet fuel), and U.S. Biofuels (FAME biodiesel).

Biodiesel Production Costs

First-Generation Biodiesel Production Costs

FAME–Soybean Oil Production Costs

Feedstock costs represent most (up to 84 percent) of the total production cost of biodiesel. At the June 2010 price of soybeans ($0.3683 per pound), the net production cost per gallon of ethanol is $3.39.[135] Table 6-6 presents a breakdown of current net production costs for FAME biodiesel.

In August 2010, the average revenue for first-generation biodiesel was $3.30 per gallon, yielding an overall $0.09 loss per gallon, including fixed costs ($0.17 return per gallon not including fixed costs).[136] Biodiesel revenue is maintained by the $1.00 tax credit per gallon, which is crucial to supporting biodiesel profitability.

Renewable Diesel Production Costs

Similar to the FAME process, feedstock costs represent most (78 percent) of the total production cost of renewable diesel. At a projected price of $0.23 per pound for yellow grease, the net production cost per gallon of renewable diesel is $2.42.[137] Table 6-7 shows a breakdown of the production costs of renewable diesel from yellow grease.[138]

Table 6-6. Breakdown of Current Net Production Costs for FAME Biodiesel

Production cost component	Production cost ($/gallon)
Corn feedstock	2.76
Natural gas	0.04
Methanol	0.11
Other variable costs	0.25
Fixed costs	0.26
Total production costs	3.42
Sales of byproducts (glycerine)	0.03
Total net production cost	3.39

Table 6-7. Breakdown of Projected Production Costs for Renewable Diesel

Production cost component	Production cost ($/gallon)
Yellow grease feedstock	1.90
Hydrogen	0.17
Other variable costs	0.08
Fixed costs	0.27
Total production costs	2.42

Second-Generation Biodiesel Production Costs

EPA has estimated commercial-scale production costs for cellulosic diesel produced from the thermochemical process (Fischer-Tropsch) from wood wastes. Table 6-8 shows the breakdown of projected cellulosic diesel costs, which are competitive with projected diesel costs.

Cellulosic diesel production requires higher initial capital investment for biorefinery construction than first-generation biodiesel. As the technology develops, these investment requirements should decrease. Average capital costs for a 33.2 million gallon capacity cellulosic diesel plant are projected at $346 million,[139] or three to four times the capital investment costs for non-cellulosic alternatives.

Third-Generation (Algal) Biodiesel Production Costs

According to EPA, algae oil feedstock costs are projected to reach $0.58 per pound, supporting a fuel production cost of $4.52 per gallon. However, estimates of these costs vary greatly: various literature sources forecast costs between $2.50 and $20.00 per gallon.

Biodiesel Tax Credit and Implications

In 2004, the American Jobs Creation Act of 2004 (Public Law 108-357) established biodiesel producer tax credits of $1.00 per gallon for "agri-biodiesel" (biodiesel produced from agricultural products, such as soybean oil or animal fats) or $0.50 per gallon for biodiesel produced from other sources (such as recycled vegetable oil). These tax credits were extended and expanded by the Emergency Economic Stabilization Act of 2008, which also increased the credit to $1.00 for all biodiesel and extended application to biodiesel made from recycled vegetable oils or animals fats, as well as renewable diesel (such as diesel derived from biomass). Although this tax credit expired on December 31, 2009, Congress may choose to reexamine this incentive in 2010 or 2011.

First-Generation Biodiesel Production Tax Credits

With a tax credit of $1.00 per gallon, first-generation biodiesel remains somewhat competitive with diesel. The tax credit is provided directly to biodiesel blenders, and it is implied in the comparison of production costs with diesel. As shown in Figure 6-6, the production cost of biodiesel rises and falls with the price of soybean oil. The current implied biodiesel production price of $2.39 per gallon (with tax credit) is equivalent to diesel production with crude oil at roughly $95 per barrel. Without the tax credit, the equivalent crude oil price would jump to $136 per barrel.

The Small Agri-Biodiesel Producer Credit provides an additional tax credit for small biodiesel production (annual capacity less than 60 million gallons) of $0.10 per gallon on the first 15 million gallons of biodiesel produced in a tax year. This credit also expired in December 2009.

Second-Generation Cellulosic Diesel Production Tax Credits

In May 2008, the Food, Conservation, and Energy Act of 2008 (Public Law 110234, Section 15321) established the CBPTC. The CBPTC provides a tax credit of up to $1.01 per gallon for producers of cellulosic diesel. This act is set to expire on December 31, 2012, but the tax credit is intended to support the early growth of the cellulosic biofuels industry.

Table 6-8. Breakdown of Projected Biochemical Cellulosic Diesel Production Costs (Wood Wastes)

Production cost component	Production cost ($/gallon)	Percentage
Biomass	1.17–1.56	40–47
Feed handling	Included above	
Other raw materials	0.05	2
Waste disposal and catalyst	0.05	2
Capital costs	1.14	34–39
Fixed costs	0.54	16–18
Total production costs	2.95–3.34	100
Naphtha and wax revenue	0.97	
Total net production cost	1.98–2.37	

Source: EPA, *Renewable Fuel Standard Program (RFS2) Regulatory Impact Analysis*, EPA420-R-10-006, February 2010.

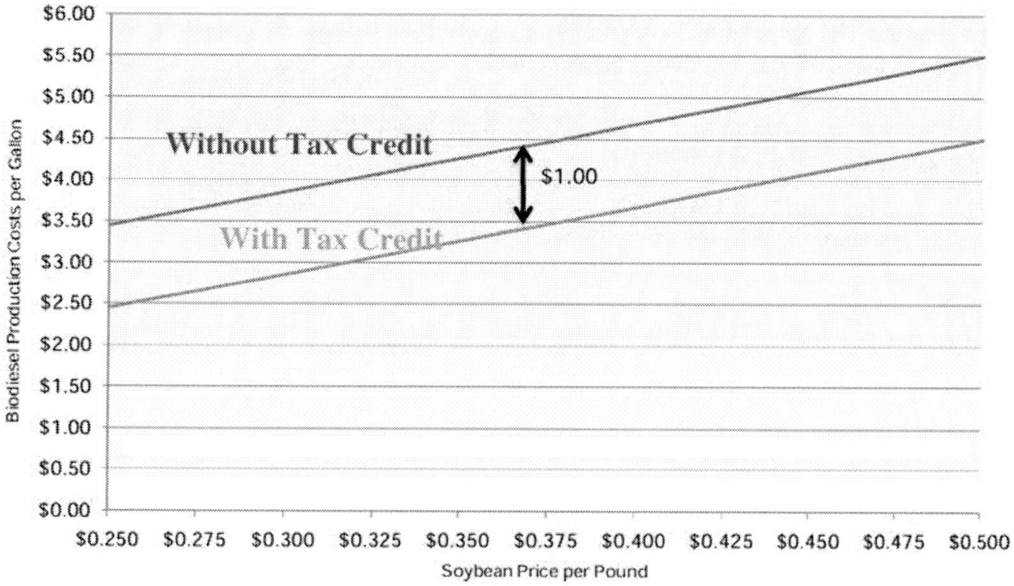

Figure 6-6. Net Biodiesel Production Costs and Commodity Price of Soybean Oil.

Future of Biodiesel, Renewable, and Cellulosic Diesel Tax Credits

In May 2010, H.R. 4213, The American Jobs and Closing Tax Loopholes Act of 2010, was introduced, which would extend the $1.00 per gallon production tax credit for biodiesel and the Small Agri-Biodiesel Producer Credit of $0.10 per gallon through December 31, 2010. In July 2010, the U.S. House Committee on Ways and Means introduced The Domestic, Manufacturing and Energy Jobs Act of 2010, which would reinstate the biodiesel tax incentives for 2011. If these credits are not extended, the viability of the domestic biodiesel industry will be in question.

Biodiesel and Renewable and Cellulosic Diesel Transportation

FAME Biodiesel Transportation

Transportation of FAME biodiesel encounters issues similar to that for ethanol— rail tank cars and tanker trucks have limited availability, and a pipeline transportation alternative is lacking. The lower production and distribution volumes for biodiesel compared with ethanol compound the transportation issues in that diseconomies of scale are accentuated. However, biodiesel production and point of sale are generally more dispersed than for ethanol, which could help support faster development of transportation infrastructure.

FAME biodiesel distribution is complicated by storage challenges, blending limitations, and physical state issues. (These issues do not apply to renewable and cellulosic diesel.) Pure biodiesel (B100) is believed to degrade to below acceptable quality if stored for periods of longer than 6 months.[140] Biodiesel can also dissolve accumulated sediments in storage and engine fuel tanks, so tanks must be cleaned thoroughly before biodiesel is added. These dissolved sediments could cause eventual fuel injection failure.[141] The National Biodiesel Board (NBB) recommends that "B100 be shipped in a way that does not lead to contamination. The association says trucks and/or railcars should be washed out before being loaded—and the only residual that is acceptable in a tanker is petroleum diesel."[142]

Biodiesel can be blended with petroleum diesel at any concentration to produce a biodiesel blend, but specific blends are necessary to meet existing ASTM fuel specifications. At blends of 5 percent and lower (B2 and B5), the fuel can be treated as fungible, conventional diesel per ASTM D975. Biodiesel blends between 6 and 20 percent fall under the ASTM standard D7467 and are marketed as commercial biodiesel.

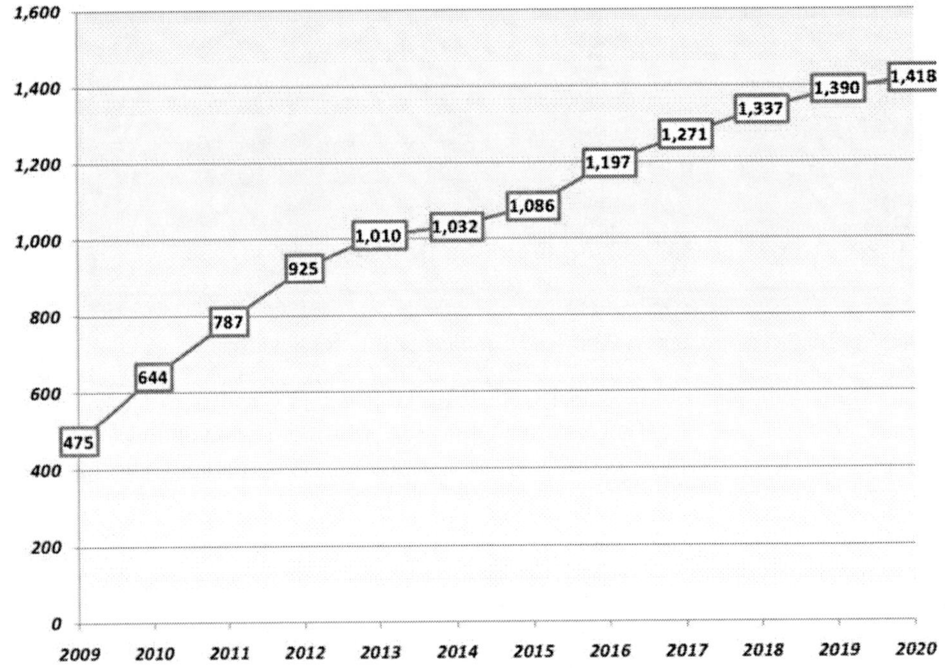

Note: Fuel quantities represent "neat," rather than blended, fuel.

Figure 6-7. Forecast of FAME Biodiesel Production and Retail Sales, 2009–20 (million gallons).

Biodiesel can be blended by one of following primary processes:

- B100 (100 percent biodiesel) splash blended with diesel fuel by the end user
- Blended by a jobber or distribution company and offered for sale as a finished blend
- Blended at a petroleum terminal or rack by a pipeline or terminal company and offered as a finished blend directly to customers or to petroleum jobbers or distribution companies for further sale to customers.[143]

Renewable and Cellulosic Diesel Transportation

The key advantage of renewable and cellulosic diesel (over FAME biodiesel) is that these drop-in diesel fuels are chemically similar to petroleum diesel and fungible within the current petroleum supply system. As such, they can use the existing transportation supply infrastructure and do not carry the same transportation and other supply chain burdens as FAME biodiesel. Renewable and cellulosic diesel can be transported through the same pipelines, use the same refueling infrastructure, and be used in the same vehicles as conventional, petroleum diesel fuel.

Biodiesel Retail Sales

FAME Biodiesel Retail Sales

Although biodiesel retail sales have grown significantly in the past 2 decades, consumer concerns over the image of diesel as a dirty fuel, performance issues in cold weather, and uncertainty over biodiesel's impact on engine warranties have hindered growth. Ultra-low sulfur diesel (ULSD) regulations, passed in October 2006, may help to combat the negative image of diesel, and biodiesel's lubricity characteristics may encourage the use of the fuel as an additive to correct ULSD's low lubricity.

We forecast production and sales of FAME biodiesel on the basis of FAPRI data. As shown in Figure 6-7 and Table 6-9, domestic FAME biodiesel production and sales are forecast to increase nearly threefold between 2009 and 2020, from 475 to 1,418 million gallons. Most of this biodiesel will be blended and sold as B20.

DOE estimates that B20 is available today at approximately 652 (0.4 percent) of the roughly 170,000 fuel stations in the United States.[144] Unlike E85 stations, which are concentrated near production facilities, B20 stations are more dispersed throughout the United States.

Biodiesel Retail Prices

B20 is competitively priced compared with diesel. Unlike ethanol, B20 has comparably high energy content to diesel (98.2 percent). In April 2010, the average price of B20 at the pump was \$3.12 per gallon (or \$2.85 per GGE).[145] This price represents a 3 percent premium over the per gallon price of diesel (\$3.02) and 5 percent premium over the per GGE diesel price (\$2.71).[146]

Renewable Diesel Retail Sales

Because renewable diesel fuel is fungible with the existing diesel supply, the only limitation of sales will likely be cost-effective production volumes. Using EPA data, we forecast that domestic renewable diesel production and sales will increase to 150 million gallons by 2020 (Figure 6-8).

Cellulosic Diesel Retail Sales

Cellulosic diesel fuel is also fungible with existing diesel supply. As a consequence, the only limitation on its sales will likely be cost-effective production volumes. Again using EPA data, we forecast that domestic cellulosic diesel production and sales will grow to 4,280 million gallons by 2020 (Figure 6-9).[147]

Table 6-9. Forecast of FAME Biodiesel Production and Retail Sales, 2009–20 (million gallons)

	2009	2010	2011	2012	2013	2014	2015	2016	2017	2018	2019	2020
Vegetable oils	434	598	724	846	914	914	956	1,037	1,081	1,107	1,120	1,108
Fats and greases	41	46	63	79	96	118	130	150	170	190	210	230
Algae	0	0	0	0	0	0	0	10	20	40	60	80
Total FAME biodiesel production	475	644	787	925	1,010	1,032	1,086	1,197	1,271	1,337	1,390	1,418

Note: Fuel quantities represent "neat," rather than blended, fuel.

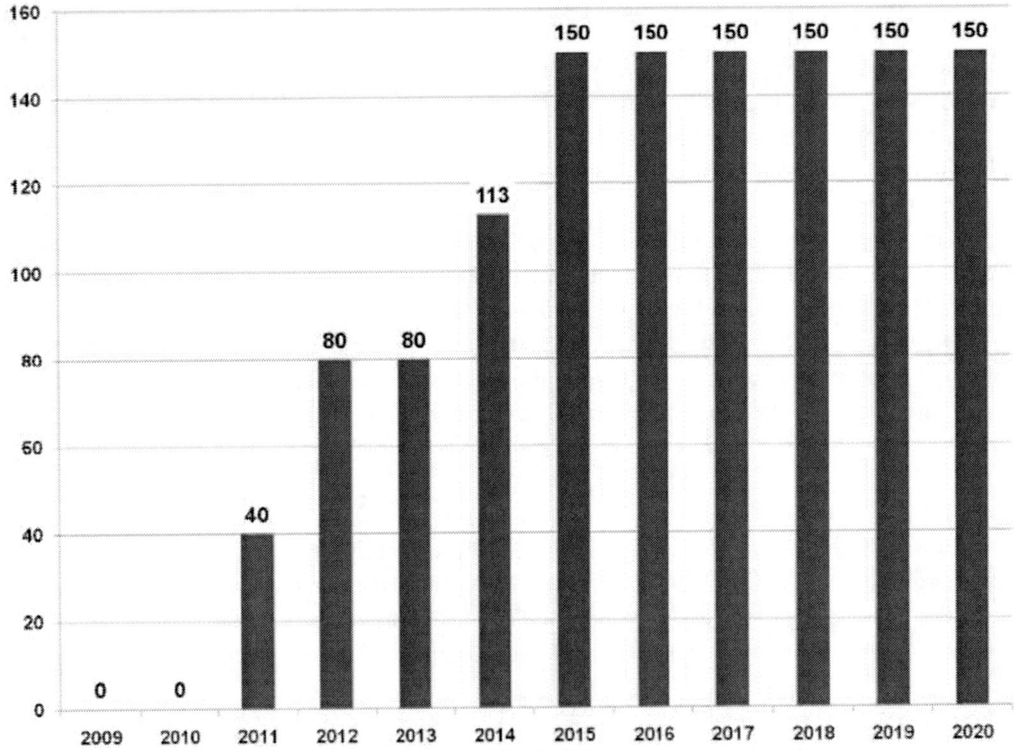

Note: Fuel quantities represent "neat," rather than blended, fuel.

Figure 6-8. Forecast of Renewable Diesel Production and Retail Sales, 2009–20 (million gallons).

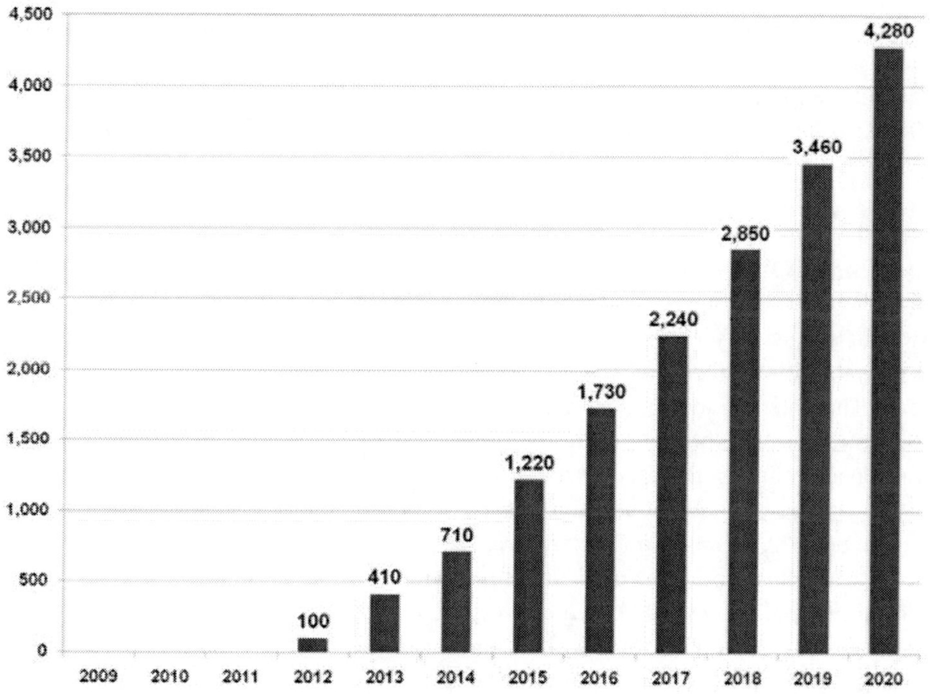

Note: Fuel quantities represent "neat," rather than blended, fuel.

Figure 6-9. Forecast of Cellulosic Diesel Production and Retail Sales, 2009–20 (million gallons).

7. Hydrotreated Renewable Jet Fuel Market Projections

In this chapter, we present our assessment of the current and future commercial availability of HRJ fuel that is not produced as an incidental coproduct of renewable diesel production. Although the feedstocks, production, and fungibility are nearly identical to those for renewable diesel, HRJ is a drop-in jet fuel (it meets the forthcoming ASTM addendum standards for HRJ)[148] that is not included in the production mandates of RFS2. Like renewable diesel, HRJ is produced from triglyceride feedstocks, such as vegetable oils (soy and canola, for example), animal fats and greases (such as tallow), plant oils (such as camelina and jatropha), and algal oils. Its oil feedstocks are converted via a hydrotreat process to produce neat HRJ, known as renewable synthetic paraffinic kerosene (bio-SPK). The neat bioSPK product is mixed in a 50/50 blend with petroleum jet fuel to produce the onspecification HRJ fuel. Although a chemically similar synthetic fuel, it is produced differently than cellulosic jet fuel, which is made from second-generation cellulosic feedstocks using the F-T process and is generally considered a secondary coproduct of cellulosic diesel.[149]

Given limited U.S. growth potential for jatropha and algal feedstock immaturity, we focus here specifically on camelina oil as it is (1) not currently included under RFS2 by EPA, and (2) considered a promising regional candidate for early commercial adoption in the U.S. Northwest. This feedstock has fast growth properties (both in cooler climates and dry land regions), can be grown on marginal agricultural land, and does not compete with other biofuel feedstocks. The estimates in this chapter are in addition to the limited volumes of HRJ potentially generated as a coproduct of renewable diesel production (Chapter 6).

Camelina appears to be a promising renewable feedstock for producing HRJ fuel, but only 68-98 million gallons of annual production capacity for camelina-based HRJ is projected by 2020.

Supply Chain Overview

HRJ fuels must rely on a complex and multi-staged supply chain. They can be derived from renewable diesel feedstocks as a fuel coproduct, including rapidly emerging camelina feedstock. The potential increase for camelina agricultural expansion is very large. In fact, by 2020 6.6 million acres throughout the Northwest could be devoted to camelina production. HRJ fuel availability is forecast to increase significantly by 2020. The increase could be driven by commercial aviation efforts, such as Commercial Aviation Alternative Fuels Initiative and the ATA and DoD alternative fuel program initiatives to increase renewable fuel usage. Camelina crops are anticipated to provide 68 million gallons of feedstock by 2015 and 274 million gallons by 2020, respectively. Estimates of initial HRJ production suggests that oil supply competition and cost will be limiting factors. Given algae oil production limits and the small range available for jathropa (Florida only), the camelina oil supply would be the only new renewable fuel feedstock that could significantly contribute to DoD's drop-in renewable fuel needs to meet its stated goals. After harvest, camelina oil must be extracted and then transported to, hydrotreated, and refined at a processing facility.

Camelina Feedstock

Native to the Mediterranean and Central Asia, *Camelina sativa,* gold of pleasure, linseed dodder, or false flax is an oil-seed plant that belongs to the mustard family.[150] With both annual and winter annual cultivar varieties, camelina is a fastgrowing crop that can thrive in cooler climates and in dry-land regions. Recent studies have identified it as a promising crop because it can germinate in nearfreezing temperatures, can grow well with limited rainfall and no irrigation, and thrives in marginal or abandoned croplands. Camelina seeds can be used to create oil that can now be processed into third-generation, drop-in replacement for petroleum-based jet fuels, known as HRJ or bio-SPK.[151] In addition to this biofuel and its associated bio-lubricant uses, camelina seed, oils, and meal contain significant amounts of omega-3 fatty acid (34 to 36 percent), known to have positive health effects, which can make it an attractive animal feed to boost omega-3 in beef, poultry, and dairy products.

Because USDA does not consider camelina a commodity crop, its planted acreage, amount harvested, and production yields are not systematically tracked. Currently, the only camelina crop statistics are from Montana (Table 7-1).

Even with increased cultivation interest in other Northwestern states, the current camelina harvests are far too low to sustain commercially viable, large-scale HRJ fuel production. Nonetheless, commercial and military aviation demand signals represent drivers for increasing investment in camelina cultivation, seed production, and oil processing plants capacity. Those efforts could reasonably be expected to increase its crop coverage within the increasingly established wheat summer fallow cropping system.

In its report *Camelina Aviation Biofuels*, Biomass Advisors projects camelina cultivation through 2020. This analysis assumes that "early adopters" will build on wheat cropping systems in the Northwest and northern Great Plains regions. Given past and current extension efforts, Montana, Oregon, Washington, and Idaho represent likely first adopters (EAG1) during the period of 2010 to 2020. Colorado, Wyoming, North Dakota, and South Dakota are also potential "early adopter" group 2 (EAG2) states, although they are initially likely to have some limited oilseed (e.g., flax and canola) experience and infrastructure, but could begin significant camelina adoption during the period of 2015–20. However, it is important to note that agricultural crops do not exist in market vacuums, so competing crops must also be considered in terms of their impacts on land and oil extraction capacities.

Figure 7-1 shows a 10-year forecast of camelina oil production, broken out by sources of land. Summer fallow is expected to supply over 50 percent of this future production. These numbers should be considered additive to the biomassdiesel forecasts until EPA approves the petition to include camelina as an acceptable RFS2 feedstock.

Aside from the petition for EPA to include it as an RFS2 feedstock, camelina has additional technical barriers and opportunities. First, camelina seeds are small, making their transportation and crushing more difficult than relatively larger seeds. Second, the resulting camelina oil can often be thick and gummy at warm temperatures, while the residual camelina meal can be prone to spontaneous combustion. Both factors could contribute to higher storage and handling costs before and during crushing. Third, the designation of camelina as a commercial commodity product is needed to support market expansion and planning. If this designation was made, the commodity markets would aid in expanding camelina cultivation acreage because producers would have an established market to sell. Finally, other beneficial actions include herbicide use registration, approval of camelina meal for animal feed, and establishing a grower support network.

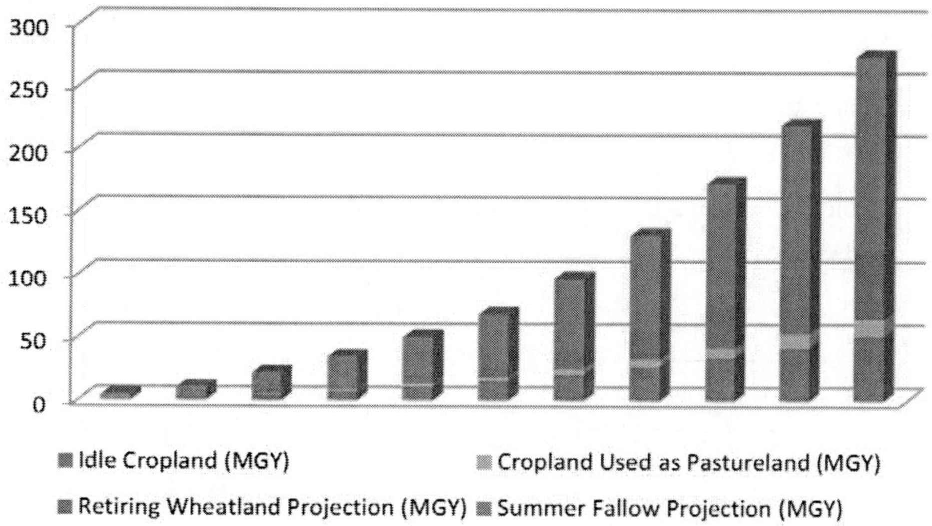

Figure 7-1. Forecast of U.S. Domestic Camelina Feedstock Oil Production, 2010–20 (million gallons).

Table 7-1. Montana Camelina Cultivation and Production, 2007–09

Camelina	2007	2008	2009
Planted	22,500	12,200	20,000
Harvested	20,400	9,100	19,500
Yield per acre	598	569	615
Pounds produced	12,197,000	5,181,500	11,998,000

Production

HRJ fuel production involves two main processes: hydrotreat processing and "biocrude" refinement. After camelina oil or other applicable feedstock oil has been extracted from the seeds, it is ready for fuel production via hydrotreatment and refining of the resultant synthetic biocrude. As a drop-in fuel, HRJ would be compatible with existing aviation infrastructure and fleets, requiring no retrofitting equipment. The drop-in jet fuel production technique involves preliminary hydroprocessing, where hydrotreatment deoxygenates the oil, followed by hydroisomerization, a catalytic process that creates hydrocarbons capable of distillation into jet fuel. To enter the market for aviation fuels, camelina-derived jet fuel must meet the same quality criteria as petroleum-based jet fuels (in accordance with the pending ASTM addendum proposal). Generally, these criteria include high energy density, high flashpoint, high thermal stability, and low freezing point.

Transportation

HRJ is a drop-in renewable fuel that is fungible with standard jet fuel and its transportation infrastructure. HRJ's fungibility with standard JET-A, JP-8, and JP5 is a key advantage for its adoption and use. To date, the Air Force, Army, and Navy have performed testing on HRJ-5 and HRJ-8 fuels derived from tallow, camelina, and algal oil. This experience and testing have demonstrated the compatibility of these fuels with existing petroleum-based jet fuel infrastructure.

Table 7-2. Average Historical Production of Montana Oilseeds, 2003–06

Oilseed	Average price ($/lb)	Total production (lb millions)	Total average value of production ($ millions)	Gross value of oil ($/gallon)[a]
Flax	0.089	26.670	2.89	1.97
Canola	0.104	20.373	2.11	1.66
Mustard	0.139	7.625	1.06	3.20
Safflower	0.132	26.745	3.53	2.38
Sunflower	0.125	4.284	0.54	1.50
Camelina[b]	0.092	12.197	1.12	2.10

Source: National Center for Appropriate Technologies, data from Montana 2006 Agricultural Statistics.

[a] Excludes pressing costs.

[b] Data available for 2007 only.

Blending

The current DoD and commercial HRJ testing, qualification, and certification efforts have focused on a blended mix of 50/50 HRJ and conventional jet fuel. These efforts are ongoing, but they have demonstrated that a maximum of a 50 percent blend is effectively a drop-in fuel with no observable safety, operational, or infrastructure difference. The technical confidence level developed with HRJs is a primary driver in the rapid movement through the original equipment manufacturer approvals and ASTM addendum balloting processes.

Costs

Several factors contribute to the economics of camelina oil and camelina-based jet fuel. These factors include costs and availability of agricultural inputs, costs of agricultural competitors, costs of production and processing relative to alternatives, subsidies for renewable fuels, and domestic and international markets for petroleum-based and renewable aviation fuels. Table 7-2 shows recent historical oilseed production and prices in Montana to provide context for feedstock impacts on neat HRJ fuel cost.[152]

Camelina production is expected to increase due to several cost considerations. First, camelina is a low-input crop that produces high yields and has a relatively low revenue positive threshold compared with other crops. Second, camelina contains high levels of omega-3 fatty acids, making it an option for dietary supplement production and a useful feed for niche omega-3 poultry, beef, and dairy products, an attractive coproduct that could help subsidize the value proposition for its cultivation. Some actions that could support expanded, cost-effective camelina cultivation acreage include commodity markets (places to sell), herbicide use registration, camelina meal use approval for animal feed, and grower support network.

Aside from these specific feedstock oil cost considerations, the ultimate HRJ costs are also coupled with those of renewable diesel production. Refineries will face a market prices tension to "tune" and produce a great proportion of renewable diesel than HRJ. This production pressure is consistent with and reflected by the conventional petroleum market dynamics (see Chapter 6).

Camelina-Based HRJ Projections

As shown in Figure 7-2 and Table 7-3, the projected level of production ranges from as much as 98 million gallons in 2020 to as little as 68 million. These production volumes are in addition to HRJ produced as a byproduct of renewable diesel production.

Retail Sales

As of 2010, DLA Energy had procured HRJ for use by the Air Force, Army, and Navy:

- The Air Force funded and procured 400,000 gallons of tallowand camelina-based HRJ-8 for testing and certification. As of January 2011, the Air Force has taken delivery of 242,900 gallons of this total. In March 2010, an A-10 Thunderbolt II flew on camelina-derived HRJ-8.
- The Army funded and procured 34,950 gallons of HRJ-8. The Army took delivery of 18,450 gallons of this total at the Southwest Research Institute in September 2010. The remaining total is scheduled for delivery in early 2011.
- The Navy funded and took delivery of 40,000 gallons of HRJ-5 derived from camelina oil at Patuxent River NAS, and in April 2010, a Navy F-18 flew on camelina-derived HRJ-5. Patuxent River NAS also took delivery of 1,500 gallons of algae-based HRJ-5 for testing and certification. The Navy funded and procured an additional 150,000 gallons of HRJ-5 in June 2010 but has not taken delivery of this fuel.

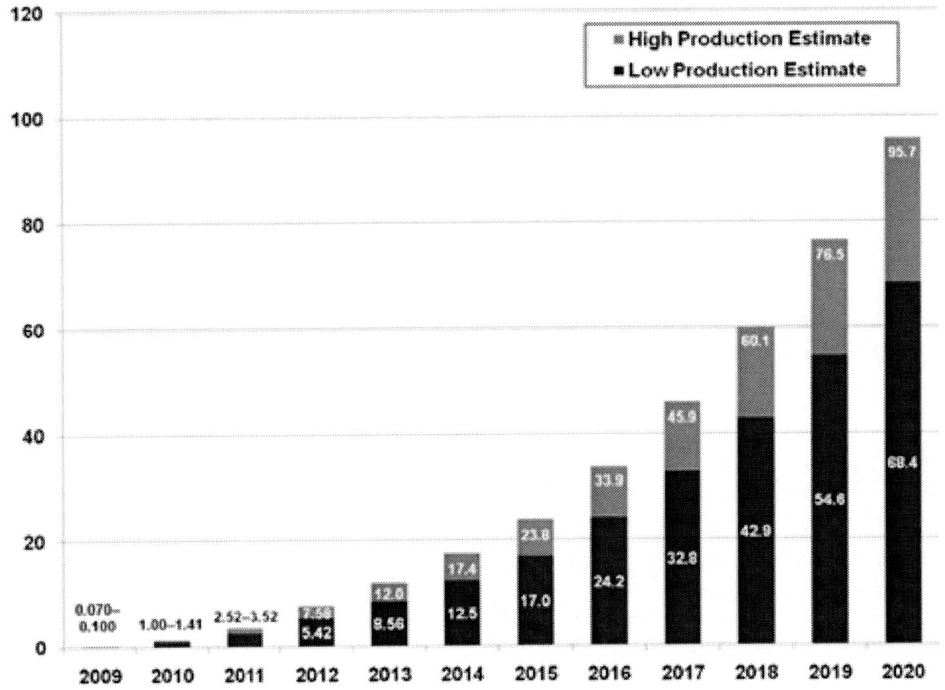

Figure 7-2. Forecast of Potential Neat HRJ Production from Camelina, 2009–20 (million gallons).

Table 7-3. Forecast of Potential HRJ Production from Camelina, 2009–20 (millions gallons)

Stage	2009	2010	2011	2012	2013	2014	2015	2016	2017	2018	2019	2020
Feedstock availability	0.298	4.02	10.1	21.7	34.2	49.8	67.9	96.7	131.1	171.6	218.4	273.5
Production capacity (low)	0.078	1.06	2.65	5.70	9.01	13.1	17.9	25.5	34.5	45.2	57.5	72.0
Production capacity (high)	0.010	1.48	3.71	7.98	12.6	18.4	25.0	35.6	48.3	63.2	80.5	100.8
Forecasted production (low)	0.070	1.00	2.52	5.42	8.56	12.5	17.0	24.2	32.8	42.9	54.6	68.4
Forecasted production (high)	0.100	1.41	3.52	7.58	12.0	17.4	23.8	33.9	45.9	60.1	76.5	95.7

The ATA, SeaTac Airport AltAir, Sustainable Oils, Honeywell UOP, and Tesoro have all partnered to establish a vertical supply chain of HRJ fuel. They plan to establish a new facility at the Tesoro Anacortes, WA, refinery to produce HRJ. This effort has already resulted in a signed (non-binding) MOU with 14 airlines to provide 750 million gallons of aviation biofuel over 10 years. On the basis of conversations with involved parties, this effort continues to move forward to a binding business arrangement as of September 2010.[153]

8. Drop-In Renewable Fuels Procurement, Logistics Integration, and Use

In this chapter, we review projected DoD demand for renewable fuels for tactical systems and weapons platforms and their compatibility with existing logistics systems (earlier chapters covered fuels for NTVs). We first review the underlying rationale for DoD use of renewable fuels, including the possible military utility and strategic considerations. In Chapter 9, we explicitly examine whether DoD should create a separate commodity class for these fuels.

The major findings associated with this chapter:

- DoD would require over 40 percent of the total projected U.S. drop-in, renewable fuel supply (regardless of fuel type) just to meet the military Services-stated level of demand, 745 million gallons in 2020.
- Drop-in renewable jet fuel production will likely not be able to the meet the Services' goal-based demand of over 570 million gallons of neat fuel.
- Meeting the Services' renewable fuel goals could require $2.2 billion in additional estimated annual fuel costs by 2020.

Reasons for DoD Use of Renewable Fuels

Military Utility

From a tactical perspective, some drop-in renewable fuels have desirable properties, such as higher cetane values or lower freeze points than conventional fuels. For example, FT-SPK, when blended with JP-8, can exhibit higher average cetane numbers than conventional JP-8 products.[154] Cetane is one measure of combustion quality in compression ignition engines. Up to a point, higher cetane numbers indicate shorter ignition delay and, as a result, more effective fuel combustion.

Several drop-in renewable fuels burn relatively cleanly, leaving fewer residues on engine components, and reduce particulate matter emissions (soot) because of their low sulfur content. These fuels therefore may reduce engine start and maintenance issues, lowering tactical system operating costs, and help meet local air quality requirements.

Another potential tactical advantage of synthetic fuels made from renewable sources is the ability to engineer them to optimize the performance of specialty engines. Engines can be designed to utilize precisely designed renewable (synthetic) fuels, leading to highly efficient operation. However, this advantage is mainly relevant for specialty applications (hypersonic aircraft, etc.).

Finally, as renewable fuels are largely sulfur free, they can be more easily used to generate electric power from fuel cells. One of the difficulties in using JP-8 for this purpose is the potential for the sulfur in the fuel to contaminate the fuel cells. In an operational setting where delivery of hydrogen is not practical, reforming hydrogen from sulfur-free renewable fuels could provide a valuable fuel option.

National Objectives

At a strategic level, U.S. energy policy objectives include a reduction in the country's reliance on petroleum-based products to lessen dependence on OPEC, broaden strategic sources of supply, improve the balance of trade, and reduce exposure to shocks from abrupt changes in the world petroleum market. As articulated by Navy Secretary Ray Mabus, "Changing the way energy is used and produced in our country is the right thing to do. It's the right thing to do for our security, it's the right thing to do for our economy, and it's the right thing to do for our environment."[155]

Increasing DoD use of renewable fuels supports a number of national objectives related to energy. For example, EO 13514, signed by President Obama in October 2009, articulates both general and specific requirements to improve federal government efficiency, decrease dependence on fossil fuels, increase the use of renewable energy, and reduce GHG emissions. EO 13514 specifically requires improved federal government efficiency through the development of a green economy. Domestically produced alternative and renewable fuels provide U.S. citizens with jobs and strengthen the nation's economy, which is key national security component.

DoD Strategic Energy Security

The 2010 Quadrennial Defense Review specifically recognizes that DoD must address energy and climate changes because of their significance to national security and strategic mission readiness.[156] Energy security for DoD means having assured access to reliable supplies of energy and the ability to protect and deliver sufficient energy to meet operational needs.[157] The reliance of our military forces on petroleum and other fossil fuels poses national security challenges. Tightening global petroleum supplies and political instability in some oil-producing nations have made prices volatile in recent years, raising costs and making budget and acquisition decisions more difficult. These associated challenges will increase if the growing demand for energy outstrips projected oil production and refining capacity.[158]

Cost Stability

Aside from these strategic priorities, reducing DoD's exposure to the price consequences of world oil market shocks would help budget planners. The broadening of strategic sources could bring some downward pressure on world petroleum prices, helping DoD meet its budget objectives, and offer supply options not otherwise available. Although DoD likely would have access to the fuel needed during a national emergency, it would benefit by having fewer such emergencies and lesser impact when they occur.

Increased sources of renewable fuels also could provide some cost stability. Since 2005, the market price of petroleum has significantly shifted more than 15 times. "This wreaks havoc on the Navy's fuel budget," according to the Naval Sea Systems Command's representative to the Navy task force on energy. Petroleum prices rose from $33 to $147 per barrel in recent years, driving the Navy's fuel bill from $1.2 billion to $5.0 billion in a single year.[159] Using renewable fuels in place of petroleum, even in part, may help hedge against fluctuations in fuel costs by diversifying the portfolio of fuel types used and thereby potentially reducing the percentage of fuel with highly variable prices.[160]

GHG Reduction

In late 2009, DoD set a target to reduce Scope 1 and 2 GHG emissions from facilities by 34 percent from FY08 to FY20. Recognizing its potential leadership role in the federal government and its role as a test bed for new technologies, DoD chose an aggressive goal that exceeds the federal government's target of 28 percent. The use of renewable fuel in place of petroleum, even in blends, would further reduce DoD's total GHG emissions. Although tactical vehicles are excluded from GHG emission reduction targets, DoD has recognized that significant reductions can be achieved in its tactical systems and weapon platforms, as asserted in its *Strategic Sustainability Performance Plan*.[161]

Renewable Fuel Market Drivers

To forecast the commercial availability of renewable fuels, each supply chain stage must be considered, including feedstock availability, fuel production (existing capacity and projected production), and transportation of fuel (to bulk facilities for blending with conventional fuels or to a retail facility for final sale to consumers). Market conditions, costs, and bottlenecks at each of these stages of the supply chain determine the availability of renewable fuels. Mandates, incentives and subsidies, and international factors also influence the level of production and costs of these fuels. The relevance of each of these factors to specific fuels is detailed in earlier chapters: ethanol (Chapter 5), biodiesel, renewable, and cellulosic diesels (Chapter 6), and HRJ (Chapter 7).

Mandates

EPAct 2005 amended the CAA to establish an RFS program, requiring a minimum volume of renewable transportation fuel produced each year. Under the RFS program, EPA "must set a standard for each year representing the amount of renewable fuel that each refiner, blender, or importer must use, expressed as a percentage of gasoline sold or introduced into commerce."

EISA 2007 revised the RFS1 requirements, establishing new volume standards for cellulosic biofuel, biomass-based diesel, advanced biofuel, and total renewable fuel that must be used in transportation fuel each year. The revised statutory requirements, referred to as the RFS2, also include new definitions and criteria for both renewable fuels and the feedstocks used to produce them, including new GHG emission thresholds as determined by life-cycle analysis.[162] However, renewable and cellulosic jet fuels are not currently subject to the volumetric production mandates.

Feedstock and Processing Availability

Renewable fuels can be produced using many different combinations of feedstock, production processes, and blending options. Appendix A, Table A-1 compares the processes, from feedstock input, through pre-processing and processing, to blending. Feedstocks are the most costly component of renewable fuels. Renewable fuels from algae are currently in the R&D stage, so they represent one of the most costly feedstock production options.

Ethanol

Corn will continue as the feedstock for approximately 98 percent of first generation ethanol produced in the United States. It will reach the limit of corn availability of 15 billion gallons in 2019. Cellulosic ethanol will likely achieve commercial viability around 2012 and support ethanol production growth into the future. Production capacity will be a larger limiting factor than feedstock availability. However, ethanol is limited in relevance for DoD as its use can only be expanded in NTVs.

Biodiesel, Renewable, and Cellulosic Diesel

Vegetable oils will continue as the feedstock for approximately 85 percent of first generation biodiesel, and biodiesel produced from fats, waste oils, and greases will grow almost tenfold by 2020. The remainder of these feedstock oils will be processed into hydrotreated renewable diesel. Cellulosic diesel will achieve commercial viability around 2012.[163] Feedstocks for cellulosic diesel include switchgrass, energy cane, sorghum, corn stover and straw, and woody biomass. Currently, biodiesel production capacity exceeds the amount of biodiesel produced, but expanding its use by DoD is limited to NTVs fleet. However, the expanded use of renewable and cellulosic diesel in tactical systems is anticipated through 2020.

Renewable Jet Fuel

Feedstocks for renewable jet fuel mirror those that apply to renewable diesel (vegetable oils, tallow, etc.) as they concurrently are produced using the same process. However, emerging second generation feedstock oils could include camelina, jatropa, and algae oils. As jatropa cultivation is primarily limited to Florida and algae oil are only in early stages of development, camelina has the potential for large additive increases over the next decade. However, camelina is not currently included as an RFS2 feedstock or cultivated in large quantities, outside of Montana.

Economics of Renewable Fuels

Government Incentives and Subsidies

Government incentives and subsidies for producing renewable fuels and feedstocks will impact price. Incentives include the following:

- Alternative fuel infrastructure tax credit, up to 50 percent of cost, or $50,000, which expires December 2010
- VEETC, $0.45 per gallon for blending, which expires December 2010
- SEPTC and small agri-biodiesel producer credit, 0.10 per gallon for first 15 million gallons produced, which expires December 2010
- Cellulosic biofuel producer tax credit, up to $1.01 per gallon for producers, including VEETC and SEPTC, which expires December 2012

- Grants and loans, such as the advanced energy research projects grants, improved energy technology loan guarantees (up to 100 percent), which does not include R&D projects, advanced biofuel production grants and loan guarantees (a maximum of $250 million), advanced biofuel production payments (other than corn starch), biomass R&D grants, and valueadded producer grants.

A description of these incentives is available in DOE's Alternative Fuels and Advanced Vehicles Data Center.[164] The expiration schedules set by legislation could impact the market availability and price of renewable fuels.

International Market Considerations

The majority of biofuel is produced in the United States, Brazil, and Europe, mostly as ethanol—either from corn or sugar. South America provides most of the renewable fuels on the international market as United States producers mostly supply fuel for domestic use under the RFS.

A DOE study on world renewable fuels production potential lists sugar-based ethanol as the least expensive biofuel, but the price in the United States is higher than that cost because a tariff is levied on ethanol from Brazil.[165] Various subsidies help renewable fuels to compete with petroleum products but are a contentious trade issue between the subsidizing countries and countries that want to enter the market.

International oil market developments likewise impact the use of renewable fuels as the price of petroleum rises and falls. Policies to stabilize renewable fuel prices vis a vis petroleum prices have limited impacts on demand for renewable fuel in the United States.

GHG regulation in other countries also increasingly impacts international markets for renewable fuels. For example, there is a proposal in the European Union (EU) to include aviation fuels as part of the EU Emissions Trading System (ETS), which requires countries to limit GHGs but enables the trading of allowances. This proposal would assign and auction allowances to airline operators at a 3 percent reduction from historical emissions in 2012 and a 5 percent reduction by 2013.[166] However, it excludes military, humanitarian, and government flights among several other categories.

Consumer Demand for Renewable Fuels

Consumer demand for renewable fuels is limited in part by the availability of infrastructure to supply these fuels. Ethanol in the form of E85 is only available at 1.2 percent of all fuel stations in the United States, mostly close to production facilities in the Midwest. B20 is available at 0.4 percent of fuel stations, but they are dispersed throughout the United States. E85 is also limited by the availability of flex fuel capable vehicles.

The amount of renewable fuel that can be blended with conventional fuels is limited. For example, ASTM D975 allows conventional diesel to contain up to 5 percent biodiesel (B5).[167] Biodiesel consumption is complicated by warranty issues, cost, and concerns about engine and filter performance. Existing legislation allows ethanol blends up to 10 percent (E10) as fuel additive to gasoline, but this market will be saturated by 2015. However, recent regulatory developments suggest that this threshold will increase to E15 prior to the 2015 blending wall.

Production Forecast for Renewable Fuels

Table 8-1 summarizes the production forecasts for renewable fuels that were presented in detail in Chapters 5 through 7. While production of all fuels is projected to increase from 2010 through 2020, the largest increases are anticipated in second and third generation feedstock fuels. Conventional ethanol, used primarily in NTV applications, will continue to comprise a large proportion of the overall quantity of renewable fuel produced in the United States and is projected to increase by 55 percent. Production of cellulosic ethanol and biodiesel are projected to increase markedly, with cellulosic ethanol comprising 17 percent of the renewable ethanol produced in 2020 and cellulosic diesel 73 percent of the biodiesel and cellulosic diesel total. Production of neat fuels suitable for 50/50 blending and use in tactical vehicles (i.e., renewable diesel, cellulosic diesel, and HRJ) is projected to increase from approximately 1 million gallons in 2010 to nearly 4.5 billion gallons by 2020, largely because of increases associated with cellulosic diesel production. Although production of renewable jet fuels is projected to increase, quantities of these fuels will remain in limited supply.

Service Renewable Fuel Goals

All of the military Services include increased use of renewable fuels in their energy strategies. In this section, we review the goals of the Army, Air Force, and Navy.

Army

In 2009, the Army released the AESIS. One of the goals of this strategy is to increase the use of renewable or alternative energy. Within that goal, the Army also set the objective of transitioning from fossil-fuel-based tactical mobility and power generation to renewable and alternative energy sources. The AESIS set goals for evaluating the performance of such fuels or blends in those systems: evaluate 50 percent of tactical ground systems by 2012; evaluate 100 percent of tactical grounds systems by 2014; evaluate 50 percent of Army aviation systems by 2014; and evaluate 100 percent of Army aviation systems by 2016.[168]

In September 2010, the Army released the Tactical Fuel and Energy Implementation Plan. The plan, pending approval, set three quantitative goals for renewable fuel: to meet 15 percent of the training base's fuel requirement with alternative fuel blends by FY18, 30 percent by FY23, and 50 percent by FY28. Using the Army's peacetime petroleum use in tactical or combat vehicles as representative of the training bases' annual fuel requirement (approximately 214 million gallons), the Army is projected to demand 16 million gallons of unblended renewable fuel in FY18, 19 million gallons in 2019, and 22 million gallons in FY20.[169]

Air Force

The Air Force has set a goal of meeting 50 percent of its CONUS fuel needs through 50/50 blends by 2016.[170] This goal implies that one-quarter of its domestic fuel needs would be met by alternative fuels in that year. The Air Force has already certified its weapons platforms, support equipment, and infrastructure to use alternative 50/50 Fischer-Tropsch synthetic/petroleum blends that can be made from renewable or non-renewable sources.

Table 8-1. Renewable Fuel Production Projections, 2010–20 (million gallons)

Fuel	Type/feedstock	2010	2011	2012	2013	2014	2015	2016	2017	2018	2019	2020
Ethanol	*Conventional Ethanol*											
	Corn	11,810	12,430	12,910	13,395	13,870	14,140	14,400	14,600	14,800	15,000	15,075
	Others	248	278	297	307	315	329	341	346	350	352	355
	Total	12,058	12,708	13,207	13,702	14,185	14,469	14,741	14,946	15,150	15,352	15,430
	Cellulosic Ethanol											
	Cellulose[a]	5	7.3	150	310	540	920	1,310	1,690	2,150	2,610	3,230
	Total	12,063	12,715	13,358	14,012	14,727	15,387	16,057	16,639	17,301	17,962	18,660
Biodiesel and Cellulosic Diesel	*FAME Biodiesel*											
	Vegetable oils[a]	598	724	846	914	914	956	1,037	1,081	1,107	1,120	1,108
	Fats and greases[a,b]	46	63	79	96	118	130	150	170	190	210	230
	Algae[a]	0	0	0	0	0	0	10	20	40	60	80
	Total	644	787	925	1,010	1,032	1,086	1,197	1,271	1,337	1,390	1,418
	Renewable Biodiesel[c]											
	Fats and greases[a]	0	40	80	80	113	150	150	150	150	150	150
	Cellulosic Dieselc											
	Cellulose[a]	0	0	100	410	710	1,220	1,730	2,240	2,850	3,460	4,280
	Total	644	827	1,105	1,500	1,850	2,456	3,077	3,661	4,337	5,000	5,848
HRJ	*HRJ from Camelina*[c]											
	Low est.	1.00	2.52	5.42	8.56	12.5	17.0	24.2	32.8	42.9	54.6	68.4
	High est.	1.41	3.52	7.58	12.0	17.4	23.8	33.9	45.9	60.1	76.5	95.7

[a] EPA, *Renewable Fuel Standard Program (RFS2) Regulatory Impact Analysis*, EPA-420-R-10-006, February 2010.
[b] FAPRI, University of Missouri, *U.S. and World Agricultural Outlook*, 2010.
[c] Neat fuel product suitable for blending and use in DoD tactical vehicles.

Table 8-2. DoD Total Tactical Renewable Fuel Consumption by Year (million gallons)

Projected demand	2010	2011	2012	2013	2014	2015	2016	2017	2018	2019	2020
Army	—	—	—	—	—	—	—	—	16.05	19.26	22.47
Air Force	0.40	0.40	0.40	0.40	129.43	258.47	387.50	387.50	387.50	387.50	387.50
Navy	0.19	0.32	0.45	0.45	0.56	1.12	1.68	84.00	168.00	252.00	336.00
Total DoD	0.59	0.72	0.85	0.85	129.99	259.59	389.18	471.50	571.55	658.76	745.97

In addition, it has set the following goal: certify the use of 50/50 HRJ renewable and petroleum fuel blends in all weapon platforms by the end of 2011. Achieving this goal will give the Air Force wider latitude in selection of fuels to meet its 2016 objective. The underlying question, however, is whether a sufficient costcompetitive supply will be available to support the achievement of this goal.

Navy

The Navy released its renewable fuel goals in October 2010. The *Energy Program for Security and Independence* and *A Navy Energy Vision for the 21st Century* outline the Navy's energy vision and strategic roadmap to achieve the Secretary of the Navy's five energy goals. These goals focus on increasing alternative fuel use ashore and afloat, greening the naval fleet, reducing petroleum use in NTVs, and purchasing and funding energy efficient systems and buildings. Specific to renewable fuels, the Navy has set the following goals:

- By 2015, the Navy will reduce petroleum use in the commercial vehicle fleet by 50 percent.
- The Navy will demonstrate a Green Strike Group in local operations by 2012 and sail it by 2016. The Navy plans to demonstrate a 50/50 blend in one or two ships by 2012, using 100,000 gallons of neat renewable jet fuel (for blended HRJ-5) and 350,000 gallons of neat renewable diesel (for blended HRF-76). The Navy then plans to sail the "Great Green Fleet" in 2016 using 50/50 blends. The carrier task force is anticipated to require the use 80,000 barrels of renewable fuel (840,000 neat gallons of renewable jet fuel [for blended HRJ-5] and 840,000 neat gallons of renewable diesel [for blended HRF-76]).[171]
- By 2020, 50 percent of total Navy energy consumption will come from alternative sources. To meet the 50 percent alternative fuel use goal in ships in that year would require 336 million gallons of neat HRJ-5 and HRF-76 in addition to planned nuclear fuel use.
- By 2020, the Navy will produce at least 50 percent of shore-based energy requirements from alternative sources.

Total Projected DoD Consumption

When projected to 2020, total demand from DoD to meet its goals for drop-in renewable fuels grows from 0.59 million gallons of neat fuel product in 2010 to 745.97 million gallons in 2020. The operational fuel demand of the military Services comprises the renewable equivalents of F-76 (marine diesel) and JP-5/8 (jet fuel). The demand for jet fuels begins to increase in FY13 and constitutes more than 76 percent of the total demand for drop-in renewable fuels by 2020.

Table 8-2 outlines this growth from 2010 to 2020 by Service based on current goals.

Projected Consumption Relative to Supply

We now compare the Services' projected collective demand for renewable drop-in fuels to the supply projected earlier in this report. Figure 8-1 shows the supply/ demand balance for these renewable fuels through 2020.[172] According to the figure, total DoD demand through

2020 would constitute almost 50 percent of total supply of drop-in renewable fuels by mid-decade. In 2020 alone, DoD demand would constitute more than 44 percent of drop-in renewable supply. These proportions constitute a much higher share of production than the Services' demands for fuels in general. For example, the 44 percent figure for 2020 compares to only 1.5 percent of the NTV ethanol and biodiesel market DoD now consumes.[173]

This raises the question of how feasible is it for the Services to obtain such a high share of the available drop-in renewable fuel supply? Much depends on the price at which these fuels will be available and on the regulations in force at that time. If the price is competitive with other fuels, DoD will also be competing with airlines and other users for every gallon. An attempt by DoD to secure over 40 percent of such a market likely would increase the price of these fuels to DoD.

A similar situation would occur if GHG reduction or other regulatory requirements effectively compel airlines and others to use drop-in renewable fuels in their operations. In such a case, DoD would have difficulty securing over 40 percent of the available supply.

Alternatively, if such fuels were sold at a price premium in the marketplace and others were not compelled to use them, DoD likely would have access to as much as it wants. In that case, the major constraint on DoD's purchases would be the budgetary implications of the associated price premium.

Figures 8-2 and 8-3 show projected supply of renewable fuels that meet the specific operational renewable fuel needs overlaid with DoD's diesel and jet fuel demands, respectively. As indicated in Figure 8-2, the projected production of renewable and cellulosic diesels should be sufficient to meet the Navy's marine diesel demands, but Figure 8-3 illustrates a large projected production shortfall for DoD's primary operational fuels (JP-5/8 jet fuels).

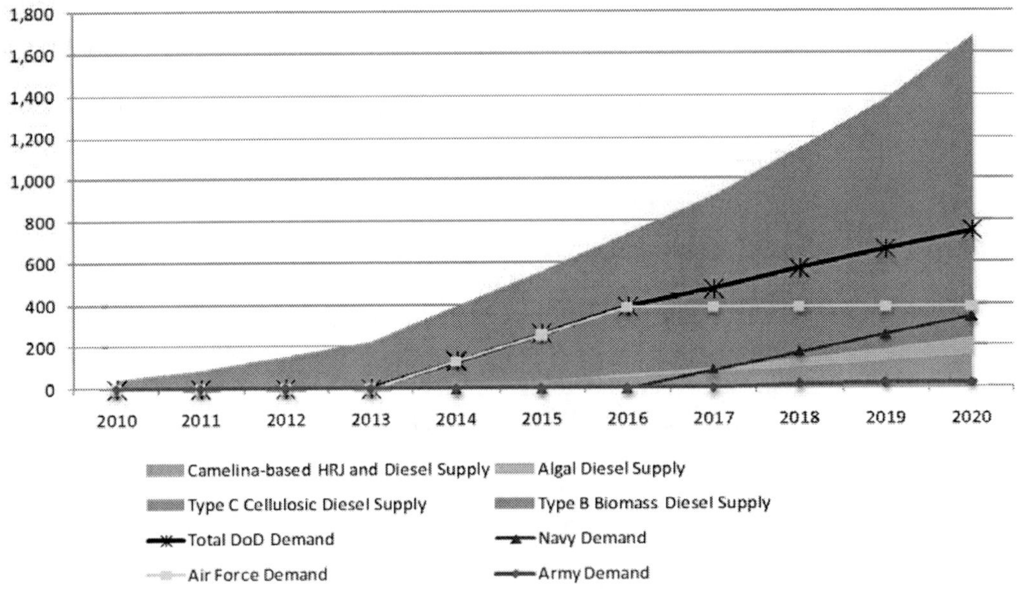

Figure 8-1. Supply/Demand Implications for DoD Drop-In Renewable Fuel Goals (million gallons).

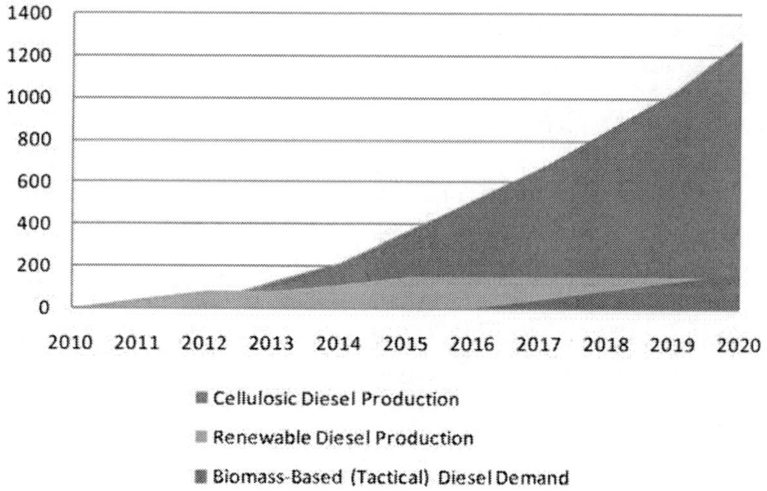

Figure 8-2. Comparison of DoD Tactical Vehicle Demand for Renewable Diesel Fuel and Projected Available Supply of These Drop-In Fuels, 2009–20 (million gallons).

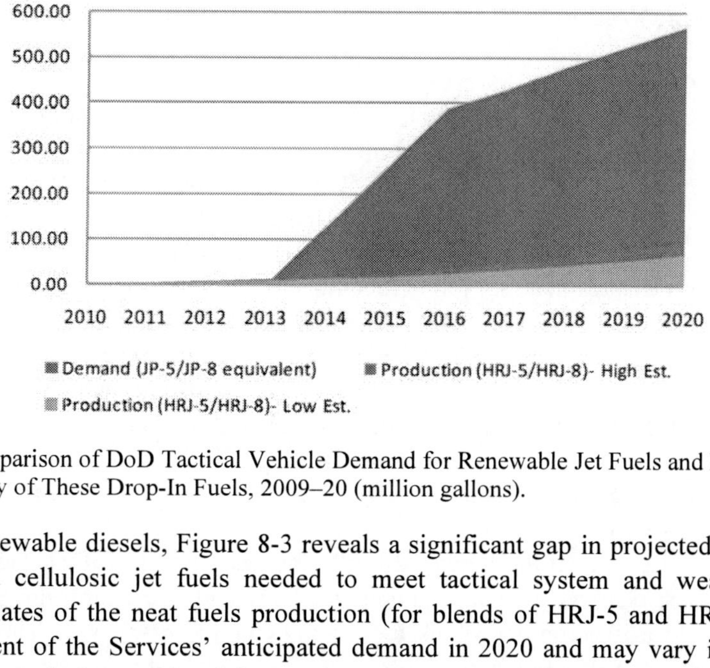

Figure 8-3. Comparison of DoD Tactical Vehicle Demand for Renewable Jet Fuels and Projected Available Supply of These Drop-In Fuels, 2009–20 (million gallons).

Unlike renewable diesels, Figure 8-3 reveals a significant gap in projected production of renewable and cellulosic jet fuels needed to meet tactical system and weapon platform demand. Estimates of the neat fuels production (for blends of HRJ-5 and HRJ-8) represent only ~20 percent of the Services' anticipated demand in 2020 and may vary in the regional availability, particularly outside of the Northwest. Some additional renewable and cellulosic jet fuel would potentially be available as coproducts of dedicated renewable and cellulosic diesel production facilities, but volumetric mandates for jet fuel are not currently included in RFS2.

Cost Implications

Currently, renewable jet fuels cost considerably more per gallon than conventional petroleum products. However, those fuels are only being produced in experimental quantities and costs would be expected to fall as production quantities increase and more is learned about how to produce them efficiently. One source reported that potential suppliers of HRJ are providing estimates to the commercial sector ranging from jet fuel wholesale market prices to $1.00 to $2.25 per gallon above wholesale.[174] Others, taking into account probable future feedstock costs and hydrotreating costs at relatively low volumes, believe the cost differential relative to conventional jet fuel would be higher.

From our analysis of feedstock and production costs (Chapters 6 and 7) and a rough estimate of the future cost of hydrotreating for a 50to 100-million-gallonper-year plant, we estimate that HRJ fuels would cost between $1.43 and $5.24 more per gallon than conventional fuels beginning in 2015.[175] As these fuels are included in a 50-50 blend with conventional petroleum, the additional cost per gallon of fuel loaded into vehicles would be half. Further, tax incentives, greaterscale economies, and technical progress could reduce the cost difference to the lower end of this range or below it. To date, however, little hard evidence indicates this will occur. In the next few years, DoD's projected consumption of dropin renewable fuels would be anticipated to represent a cost premium of $865 million dollars annually by 2015 and $2.2 billion annually by 2020, assuming a midrange projection.

Tactical System, Weapons Platform, and Infrastructure Compatibility

Infrastructure Compatibility

Before being considered drop-in fuel, these new renewable fuel products must first be tested to ensure that their properties are chemically and physically similar enough to meet the respective fuel specifications when blended (50/50) with conventional petroleum products. For example, unlike FAME biodiesel, renewable and cellulosic diesels are chemically similar to petroleum diesel. Renewable diesel must meet the ASTM D975 standard. Ongoing laboratory analysis ensures that renewable fuel indeed meets the appropriate specification but is likewise followed by extensive "fit for purpose" testing to check the fuel's real world characteristics. Preliminary information suggests that these drop-in fuels can be transported through the same pipelines, use the same refueling infrastructure, and be used in the same vehicles as diesel fuel, without causing damage or malfunction.[176] However, bench top findings alone are not sufficient to assure infrastructure compatibility. As such, renewable fuel products' performance (and compatibility) must be assessed for materials, components, and distribution systems.

Military Services have their own distinct fuel product and fuel facility technical communities. Each Service maintains unique policies and procedures for characteristics, performance, and certification of its bulk fuel supply chain infrastructure. The Services have already begun extensively testing renewable fuels' material compatibility as part of their respective fit for purpose and performance testing programs. As the DoD moves towards joint bulk fuel management and operations, Service harmonization of their policies, protocols, and procedures is increasingly important.

On December 3, 2009, the DoD Energy Task Force and the TRI-Service POL Users Group established the Alternative Fuels Harmonization Initiative. Its purpose is to focus on harmonizing alternative fuel test and certification requirements and technology development efforts across DoD. The initiative's activities include functional areas, such as: aircraft; maritime; vehicles and equipment; environmental, safety, and occupational health; and infrastructure.[177] These efforts are complemented by the TRI-Service POL Users Group's

SubCommittee on Alternative Fuels Supply Chain Infrastructure Certification. Its aim is to "integrate and coordinate service requirements and efforts for certification of DoD Class III bulk fuel supply chain infrastructure on the use of new alternative aviation and ground fuels to ensure the operational readiness/affordability of legacy and emerging systems."[178] While OASD (OEPP)'s emergence is realigning these roles, the Services' harmonization of infrastructure certification activities will remain critically important for technical risk assessment, management, and mitigation as drop-in renewable fuels are integrated into the bulk fuel supply chain.

Based upon the Services' preliminary test and certification efforts, drop-in renewable and cellulosic fuels are expected to be compatible with the existing fuel distribution infrastructure. Likewise, new fuel blending capabilities should not be required. While the preliminary technical assessments suggest little or no negative impact, the Services' technical communities will need to continue and complete their infrastructure certification processes prior to the full scale integration of drop-in renewable fuel into the bulk fuel supply chain.

Finally, introducing drop-in renewable fuels into the distribution infrastructure will still require careful fuel monitoring into the future. It will be necessary to ensure that delivered renewable product does not exceed 50 percent of the blended fuel and meets Services' fuel specifications. Furthermore, continued monitoring of operation and maintenance will require long-term observation to assess whether there are any adverse impacts to the distribution infrastructure.

Tactical System and Weapons Platform Compatibility

To be a useable military commodity, a drop-in renewable fuel must meet or exceed the appropriate fuel specification criteria, fit for purpose evaluations, performance testing, qualification, and certification prior to use in a tactical system or weapon platform. Military Services are each responsible for certain fuel specifications (Air Force is responsible for JP-8, Navy is responsible for F-76, etc.). Military fuel specifications cover such qualities as energy density, material compatibility, lubricity, flash point, freezing point, viscosity, stability (thermal and microbial), water separability, etc.

The energy density of a drop-in fuel is critically important because it affects the ranges of tactical systems and weapons platforms that use the fuel, the amount of fuel that must be carried to achieve a given range, and the logistics system needed to support the fuel (more or bigger fuel tanks are needed to support a less-energydense liquid fuel). Energy density is evaluated for both specific energy, or gravimetric energy content (mass of fuel required), and volumetric energy content (volume of fuel required).[179] Fischer-Tropsch synthetic fuels have slightly higher specific energy, but slightly lower energy density than petroleum-based fuels (indicated as Jet A and Jet A-1).[180] Drop-in renewable fuels, such as HRJ, have energy densities close to those of conventional petroleum products and hence are acceptable for use in tactical systems and weapons platforms.

From a material compatibility standpoint, hydrotreated renewable and cellulosic Fischer-Tropsch fuels do not typically contain significant amounts of aromatic compounds. While this allows them to burn more cleanly with respect to particulate emissions, it also can lead to fuel leaks because the elastomers used in the fuel system seals may shrink.[181] However, this concern largely eliminated when the synthetic fuels are blended with conventional petroleum.

These renewable and cellulosic fuels also have lesser lubrication properties than petroleum-based fuels. However, this problem may be overcome through blending or the addition of lubricants into the fuel stream.[182] Finally, renewable and cellulosic fuels can perform well at low temperatures and maintain low viscosity, enabling blends containing them to add performance capacity at high altitudes.[183]

As noted previously, Air Force testing and demonstrations of blended FischerTropsch synthetic fuels has already successfully certified its acceptability for use in many of their weapons platforms. Researchers at the AFRL indicate they expect similar results for HRJ because of the end use fuel's chemical similarity to coal-derived FT-SPK.[184] More generally, because drop-in renewable and cellulosic (synthetic) fuels can be chemically structured to mimic almost any known fuel, they can be shaped to have many of the same properties as JP-8. The main question regarding these fuels, therefore, is not whether they can be made compatible with infrastructure or weapons systems, but their producibility (can they be produced at a scale of output that is realistically feasible and cost competitive).

Building on their earlier alternative fuel testing and evaluation experience, the Services' technical communities are aggressively moving forward with their respective renewable fuel test and evaluation roadmaps. While the Services each have their own programs, they are actively working to harmonize these test, qualification, and certification efforts and to rapidly share results through various technical forums. For instance, technical exchanges are being performed in venues, such as the Aviation Working Group. These harmonization and exchange efforts have contributed to the rapid progress by each of the Services' technical communities. Table 8-3 is a high level representation of RDT&E, qualification, and certification of relevant organizations across the DoD.

Table 8-3. DoD Organizations Conducting Fuel Testing and Evaluation

Service	Organization	Activity					
		Research & development	Laboratory evaluations	Component evaluations	System evaluations	Demon-strations	Certification & specifications
		Fuel, system, or logistical support R&D	Fuel specifications, fit for use, and/or other tests	Fuel use in components, engines, and distribution systems	Small vehicle, equipment, platform, or weapon systems demos	Large field service evaluations for long term operability	Modification or revision of specifications and certification
DoD	DARPA	x	x				
	DLA Energy	x	x				
Army	TARDEC NAC	x	x	x	x	x	x
	CERDEC Ground Vehicle Power and Mobility	x		x	x		
	AMRDEC	x		x	x	x	x
Navy	NAVSEA, NAVAIR, NAVSUP, USMC, ONR	x	x	x	x	x	x
Air Force	AFRL	x	x	x			
	AFCO			x	x		x
	AFPA	x	x	x			x

Summary

DoD's military Services have made commitments to pursue drop-in renewable fuel use for reasons of national security, energy security, and GHG reduction. They have all established goals for the use of such fuels, and these quantities would constitute a fairly large portion of the likely available renewable jet fuel market in 2020. To enable the achievement of these goals, the Services are all conducting extensive testing and evaluation to ensure compatibility with existing tactical systems, weapons platforms, and infrastructure. However, the supply of drop-in renewable fuels available for procurement, at least for the next several years, and the likely cost premiums represent some significant barriers.

Substantial uncertainty surrounds the costs of production of these renewable fuels, both their absolute cost and how quickly those costs would drop with increased scale economies and technical progress. The practically available resource base for the fuels is also uncertain, though generally that supply could be expected to rise with price (the higher the price paid resource suppliers are, the more that could be obtained). Years of further experience will probably be necessary to determine whether these prices can both elicit a sufficient resource supply and still allow the fuels processed from this supply to strongly compete in the marketplace.

9. Potential Renewable Fuel Commodity Class

Section 334 of NDAA FY10 asks that the present study assess whether DoD should establish a renewable fuel commodity class. We describe what it would mean for DoD to establish a renewable fuel commodity class and how creating such a class alongside the existing Class III bulk petroleum supply chain would affect DLA fuel management. We also consider the pros and cons of creating such a new commodity class and offer conclusions regarding its advisability.

Creating a new DoD commodity class might help renewable fuels gain greater visibility in some respects, but the additional costs and potential adverse effects outweigh the potential benefits.

DoD Commodity Classes

DoD currently manages the following commodity classes:

- Class I: Subsistence
- Class II: Clothing, individual equipment, tools, administrative, and supplies
- Class III: Petroleum, oils, and lubricants (POLs)
- Class IV: Construction materiel
- Class V: Ammunition
- Class VI: Personal demand items
- Class VII: Major end items
- Class VIII: Medical materiel

- Class IX: Repair parts
- Class X: Materiel for nonmilitary programs.[185]

The question is whether DoD should add another commodity class, one that breaks out a separate fuel class from those included in Class III.

DoD military doctrine is organized to some extent around these commodity classes. Each class has a distinct role in defense planning, and each must be properly resourced and distributed to maximize operational effectiveness. A change in the composition of a commodity class, therefore, would affect DoD supply doctrine and operational planning. It also would impact DoD logistics networks and training.

DLA Fuels Management

DLA acts as DoD's agent in procuring logistics supplies, including fuels and other energy products. Specifically, DLA Energy manages the supply chain for Class III bulk petroleum and other energy commodities. In doing so, it acts not only for DoD, but also for other federal agencies, state and local governments, and authorized commercial entities.

DLA Energy currently has the following commodity business units:

- Bulk fuels
- Facilities and distribution management
- Direct delivery fuels
- Installation energy
- Energy enterprise
- Missile fuels.

DLA Energy B is the executive agent for the bulk fuel commodity class. As the executive agent, it provides worldwide support in the areas of contracting, distribution, transportation, and inventory control of bulk fuels, including jet fuels, distillate fuels, automotive gasoline (for overseas locations only), aircraft engine oils, and crude oil, in support of the DOE Strategic Petroleum Reserve Program.[186] DLA Energy B also has divisions that manage contracting, inventory and distribution, and transportation rates and policy.

If DoD established a new renewable fuel commodity class, DLA Energy would presumably create a new fuel category that it would manage separately from petroleum-based fuels. Such a category could be created by establishing a new functional office within the existing organizational structure. Thus, DLA Energy would need a seventh commodity-level business unit that focuses on renewable fuels. The purpose of the unit would be to manage the acquisition and distribution of drop-in renewable fuels separately from other bulk fuels. DLA Energy currently applies national stock numbers (NSNs) to individual fuels,[187] so the new entity would need to establish new NSNs for renewable fuels and advance other distinctions to emphasize their uniqueness within the DLA bulk fuels category.

Findings

Section 334 asks whether the national interest in renewable fuel use by DoD is compelling enough to justify the implied changes in doctrine, training, logistics networks, and DLA Energy operating procedures. The creation of a new fuel commodity class, separate from petroleum, would focus DoD leadership's attention on this commodity and make it more visible throughout the defense community. However, creating a DoD renewable fuel commodity class separate from the Class III POL supply chain involves issues that raise doubt as to the wisdom of such an action.

For one, the military Services are striving to make such fuels interchangeable with petroleum. This thrust is consistent with DoD's one-fuel-on-the-battlefield doctrine, which aims to facilitate the simplified logistics of supplying fuels for expeditionary operations. Because drop-in renewable fuels, such as HRJ-8, can be processed to approximate the properties of conventional petroleum JP-8, the Services' desire to integrate them into the existing supply chain appears achievable. Creation of a separate renewable fuel commodity class would be inconsistent with this approach and could even create additional infrastructure and management costs while deterring user acceptance.

Second, renewable fuels only offer some limited military utility. Although some of these fuels have desirable properties (discussed in Chapter 8), our review of drop-in renewable fuels indicates that their advantages over conventional military specification fuels are limited to cleaner combustion, potential for reduced engine start and maintenance issues, and usefulness in some specialty applications (lower freeze point, etc.). DoD's existing commodity classes appear to identify distinct categories of items essential to military operations. Although renewable fuels may be useful in some respects, they do not appear sufficiently distinct or superior from a military utility perspective to merit creation of a separate commodity class.

Third, many of these fuels are expensive to produce, and how rapidly their costs will drop over time is unclear. As a result, DoD could suffer adverse budget consequences from large purchases, at least over the next decade, perhaps causing a reevaluation of the commitment in the future. By implication, compelling the creation of a separate DoD commodity class appears premature.

Fourth, although DoD is an important purchaser of fuels, it captures only a small share of the overall U.S. market, so if these fuels are to be widely adopted, they must attract interest from a much larger class of buyers. The creation of a separate commodity class could encourage DoD use of these fuels, but their ultimate success depends more on whether they can compete in the commercial sphere, particularly in the commercial transportation industry.

Finally, establishing a separate renewable fuels commodity category would clearly distinguish them from other bulk fuels and demonstrate an ongoing commitment on DLA's part to securing such fuels. However, DLA responds to the Services' demands for supplies, not the other way around. In this case, the impetus for the DLA Energy focus on renewable fuels is coming from the Services' desires to begin using them in increasing amounts, not from a DLAinitiated program to supply such fuels. Thus, DLA Energy has no evident organizational need to adopt a separate renewable fuel commodity class to emphasize their acquisition.

Also, whether it makes sense to create a new business unit in DLA Energy to handle drop-in renewable fuels is unclear. Presumably, the answer lies in whether such a unit could negotiate contracts, plan transport, and perform other necessary duties more efficiently than, for example, a subunit within DLA's existing bulk fuels unit. This question is independent of whether creating a new commodity class is advisable.

Should DLA apply new NSNs to various drop-in renewable fuels? DLA currently applies a number of NSNs to petroleum fuels, so it makes sense to develop new ones for neat or pure renewable drop-ins or their blends. However, whether to do so and how many to attach can only be resolved after DLA obtains more experience purchasing and delivering these fuels.

The tracking of drop-in renewable fuel use requires attention. DLA Energy will need to track the purchase and disposition of drop-in renewable fuels to ensure that Service consumption goals are met and to calculate the budgetary consequences. This will require attention to identification methods for these fuels and to accounting procedures, probably in cooperation with DLA Energy's various fuel suppliers. However, creation of a new fuel commodity class is not necessary to accomplish these objectives.

In summary, Congress requests that DoD focus on drop-in renewable fuels and asked whether creation of a separate commodity class would be advisable. We conclude that creation of this class could help these fuels gain acceptance in the defense community, but that the associated redundant infrastructure costs and user acceptance impediments would considerably outweigh the benefits. Further, the DoD community is already strongly interested in drop-in renewable fuels and seeking to integrate them into its existing commodity structure. This integration will expedite their acceptance and minimize the burden on the logistics infrastructure. Therefore, the creation of a DoD renewable fuel commodity class is not justified or beneficial overall.

CONCLUSION

In this chapter, we present our conclusions regarding DoD use and potential use of renewable fuels. We also look at supportive measures DoD, the Executive Branch, and Congress can take to ensure an available supply of these fuels. (In Chapter 9, we offered our conclusion regarding the creation of a separate commodity class for these fuels.)

DOD Use of Renewable Fuels

From our findings, DoD's NTV use over the last decade has been primarily driven by the Services' sustainability programs and their efforts to contribute to DoD achievement of the federal fleet alternative fuel goals. While DoD's renewable fuel consumption continues to grow, it is primarily limited to supplanting the Services' use of petroleum in their non-tactical fleet vehicles. This said, military Services' stakeholders noted that this growth has been costly in terms of new vehicles, modified infrastructure, and increased maintenance of the NTV fleet, particularly among diesel vehicles.

However, as discussed in Chapter 2, the vast majority of DoD's petroleum fuel use is in tactical systems and weapons platforms. While these represent the largest opportunity for expanded renewable fuel use, they are mission essential equipment that are appropriately subject to lower risk tolerance for performance and reliability because the lives of our Service members and our national security depend on the superior performance and reliability. Replacement of new tactical systems and weapons platforms takes place on almost a generational scale so renewable fuels must conform to existing fuel specifications and performance requirements (must be a drop-in replacement). Before their use in a tactical system, renewable fuels must be qualified and the weapons platform certified to ensure the fuel does not compromise either the mission or soldier safety. In addition, given long-term DoD efforts to simplify POL logistics (use of a single battlefield fuel, JP-8), proposed renewable fuels that require separate supply chains or are incompatible with existing infrastructure face adoption challenges. For example, use of biodiesel (FAME) has been pursued in tactical systems since the Marine Corps' early work almost a decade ago. After significant evaluation and testing, which demonstrated certain incompatibilities, the warfighter technical and logistics communities (the Tri-Service POL Users Group) chose to prohibit the use of FAME biodiesel in tactical systems. This no-use policy is agreed upon across the Services.

The DoD and military Service technical and logistical communities support—and even are aggressively moving toward—enabling the use of new drop-in renewable and cellulosic jet and marine diesel fuels. The early work of their technical communities on non-renewable synthetic fuels developed the data, experience, and process to allow the use of Fischer-Tropsch-based synthetic fuels as well as the newly emerging hydrotreated renewable fuels. All of the Services have qualification and certification road maps that will enable widespread usage of these new drop-in renewable fuels in tactical systems operating on the land, on the sea, and in the air. Renewable fuels meeting the respective ASTM and Service fuel specifications should preclude the need for separate logistics infrastructure and minimize maintenance risks to existing infrastructure and assets.

The Services have goals to enable use of drop-in renewable fuels. Some have established stretch goals to use significant amounts of the drop-in fuels to meet their tactical systems' fuel requirements out to 2020. These quantities are significant enough that the primary barriers are whether U.S. renewable and cellulosic fuel producers will be able to meet the production quantities and at a competitive cost (or at an acceptable price premium) to meet both DoD and commercial demand. Integral to these two keystone issues are the following:

- The ability to raise capital investment for hydrotreat infrastructure expansion at existing and new refineries and for the establishment of cellulosic Fischer-Tropsch biorefineries, given uncertainties about their relative costs.
- The cost-effectiveness and availability of feedstock resources for the fuels.

By procuring fuels already covered under the RFS2 fuel categories framework, DLA Energy, as DoD's executive agent, can ensure compliance with EISA Section 526 requirements. For feedstocks or renewable mobility fuel not yet included under RFS2 volumetric mandates (hydrotreated renewable and cellulosic jet fuel), DoD, as well as the U.S. commercial air transportation sector, would benefit from the expansion of RFS2 volumetric mandates to include these fuels.

Supportive Measures to Ensure Supply

The primary challenge to meeting the Services' tactical system and weapon platform renewable fuel goals include (1) ensuring sufficient drop-in renewable fuel supply and (2) that it is available at an acceptable cost. Because of the magnitude and breadth of this challenge, the options identified apply to DoD, other executive branch agencies, and Congress.

During the next 10 years, DoD could potentially positively enhance renewable fuel feedstock availability, processing, and production through the use of longterm contracting authority. If provided such a capability for renewable fuel purchases, DoD would help address the capital investment conundrum within the renewable fuel industry. Without a medium-term guaranteed demand and price floors for renewable fuels, many renewable fuel products lose their cost competitiveness (even with incentives) because of the short 5-year period that investors need to recover their capital investment costs. Although DoD has requested 20-year contract authority, similar commercial industry efforts have suggested that even 10 years would represent the tipping point for more mature renewable fuel producers to obtain financing to build the necessary infrastructure and plants. In addition, renewable fuels during the next 10 years likely will require premium pricing over conventional petroleum commodity fuels. As a result, the DoD bulk fuel contracting model may need to be reconsidered given the nature of the dropin renewable fuel market.

Other executive branch agency actions that would facilitate renewable fuel supply include collaboration and action from USDA, EPA, and DOE. For example, USDA recognition of emerging energy crops as commodities would help growers obtain crop insurance and sell their product as a commodity crop. Other USDA actions could include more extension research and support of these feedstocks. EPA's role in the execution of RFS2 makes it a key partner in fast-tracking the feedstock petition process and collaboratively providing transparent life-cycle GHG evaluations of new feedstocks and fuel pathways. DOE could continue its research and development leadership, investment in high-intensity feedstock production systems, and work with Congress to ensure strategic-loan guarantees are available for mature drop-in renewable fuel processing and production infrastructure. The relative attractiveness of these measures depends on the capital intensity of a given renewable fuel production process and the type of feedstock involved. However, the current availability of market capital suggests that all of these efforts would have some impact on federal budgets.

Another mechanism for ensuring an acceptable supply of renewable fuels for DoD tactical systems and weapons platforms is the Title III Program, which is part of the Defense Production Act (DPA). This option could be specifically utilized for developing renewable fuels infrastructure. The mission of DPA Title III is to "create assured, affordable, and commercially viable production capabilities and capacities for items essential for national defense." The Defense Production Act Reauthorization of 2009 states:

> to further assure the adequate maintenance of the domestic industrial base, to the maximum extent possible, domestic energy supplies should be augmented through reliance on r enewable energy sources (including solar, geothermal, wind, and biomass sources), more efficient energy storage and distribution technologies, and energy conservation measures.

Use of Title III could allow for the development of infrastructure as well as an industrial base, particularly for the more capital-intensive, cellulosic synthetic fuels. However, the program's focus is on technology transfer agreements and often involves selecting a "winner" to receive startup capital investment, something which could generate political criticism and may be subject to schedule risks. All projects awarded are long term, so they are most likely to produce fuels after 2020 time horizon.

Congress has the option to directly help ensure drop-in renewable fuel supply by directing EPA to add renewable jet fuel to RFS2 production mandates and establish production incentives (such as those that recently expired for biodiesel). Congress could likewise support budget additions for the development of renewable and cellulosic drop-in fuel processing infrastructure to directly spur production capacity that would likely produce "fungible HRJ" and cellulosic jet fuels in the midterm.

In conclusion, the military Services have taken several important steps including comprehensive test and evaluation programs and plans to procure fairly substantial volumes of renewable fuels over time. We found that most of DoD's tactical systems and weapons platforms will have the flexibility to utilize these fuels within CONUS or in overseas theaters by 2016. However, the projected U.S. supply of drop-in renewable fuels will not be sufficient to meet anticipated DoD demand, specifically for renewable and cellulosic jet fuel products. Additionally, price premiums for drop-in renewable fuels and the budgetary implications associated with renewable fuel goals may be prohibitive. DoD has the motivation to explore the use of renewable fuels to realize some limited tactical military utility, to support its strategic energy security goals, and to contribute to the larger U.S. national security interests. Further action by DoD, the relevant Executive Branch agencies, and Congress is needed to promote jet fuel production and address the price premiums in order to help the Services achieve their renewable fuel goals.

Appendix A. Renewable Fuel Market Drivers

In this appendix, we present an overview of the factors influencing the current and future commercial availability of renewable fuels. To forecast the commercial availability of renewable fuels, we consider each supply chain stage. For emerging renewable fuels, the supply chain includes feedstock availability, fuel production (capacity and projected production), and transportation of fuel (to bulk facilities for blending with conventional fuels or to a retail facility for final sale to consumers). Market conditions, costs, and bottlenecks at each of these stages of the supply chain determine the availability of renewable fuels. Mandates, incentives and subsidies, and international factors influencing production also are included. Each of these factors is detailed in other chapters for specific fuels: ethanol (Chapter 5), biodiesel, renewable, and cellulosic diesels (Chapter 6), HRJ (Chapter 7), and biobutanol (Appendix D).

Mandates

EPAct 2005 amended the CAA to establish an RFS program, requiring a minimum volume of renewable transportation fuel produced each year. Under the RFS program, EPA "must set a standard for each year representing the amount of renewable fuel that each refiner, blender, or importer must use, expressed as a percentage of gasoline sold or introduced into commerce."

EISA 2007 revised the RFS1 requirements, establishing new

> volume standards for cellulosic biofuel, biomass-based diesel, advanced biofuel, and total renewable fuel that must be used in transportation fuel each year. The revised statutory requirements, referred to as the RFS2, also include new definitions and criteria for both renewable fuels and the feedstocks used to produce them, including new greenhouse gas (GHG) emission thresholds as determined by life-cycle analysis.[188]

Feedstock and Processing Availability

Renewable fuels can be produced using many different combinations of feedstock, production processes (pre-processing, processing, and production), and blending options. Table A-1 compares the alternative fuel production process, from feedstock input, through pre-processing and processing, to blending. The last column indicates whether each renewable fuel is solely for NTV use or operational fuel. The processes depicted in the table vary from established, commercially viable processes to those in the early stage of R&D. Feedstocks are the most costly component of renewable fuels. Renewable fuels from algae are currently in the R&D stage, so they represent one of the most costly.

Ethanol

Corn will continue as the feedstock for approximately 98 percent of first generation ethanol produced in the United States. It will reach the limit of corn availability of 15 billion gallons in 2019. Cellulosic ethanol will likely achieve commercial viability around 2012 and support ethanol production growth into the future. Production capacity will be a larger limiting factor than feedstock availability.

Biodiesel, Renewable, and Cellulosic Diesel

Vegetable oils will continue as the feedstock for approximately 85 percent of first generation biodiesel, and biodiesel produced from fats, waste oils, and greases will grow almost tenfold by 2020. Cellulosic biodiesel will achieve commercial viability around 2012.[189] Feedstocks for cellulosic biodiesel include switchgrass, energy cane, sorghum, corn stover and straw, and woody biomass. Currently, bio-diesel production capacity exceeds the amount of biodiesel produced.

Renewable Jet Fuel

Feedstocks for renewable jet fuel mirror those that apply to renewable diesel (vegetable oils, tallow, etc.) as they concurrently are produced using the same process. However, emerging second generation feedstocks could include camelina, jathropa, and algae oils.

Although jathropa cultivation is primarily limited to Florida and algae oil are only in early stages of development, camelina has the potential for large increases over the next decade. However, it is not currently included as an RFS2 feedstock or cultivated in large quantities, outside of Montana.

Economics of Renewable Fuels

Government Incentives and Subsidies

Government incentives and subsidies for producing renewable fuels and feedstocks will impact price. Incentives include the following:

- Alternative fuel infrastructure tax credit, up to 50 percent of cost, or $50,000, which expires December 2010
- VEETC, $0.45 per gallon for blending, which expires December 2010
- SEPTC and small agri-biodiesel producer credit, $0.10 per gallon for first 15 million gallons produced, which expires December 2010
- Cellulosic biofuel producer tax credit, up to $1.01 per gallon for producers, including VEETC and SEPTC, which expires December 2012
- Grants and loans, such as the advanced energy research projects grants, improved energy technology loan guarantees (up to 100 percent), which does not include R&D projects, advanced biofuel production grants and loan guarantees (a maximum of $250 million), advanced biofuel production payments (other than corn starch), biomass R&D grants, and value-added producer grants.

A description of these incentives is available in DOE's Alternative Fuels and Advanced Vehicles Data Center.[190] The expiration schedules set by legislation could impact the market availability and price of renewable fuels. We discuss the implications of tax credits on specific renewable fuels in later sections of this report.

International Market Considerations

The majority of biofuel is produced in the United States, Brazil, and Europe, mostly as ethanol—either from corn or sugar. South America provides most of the renewable fuels on the international market as U.S. producers mostly supply fuel for domestic use under the RFS.

Government incentives enable U.S. producers to compete with producers in other countries, but they have variable impacts on efficiency. For example, Brazil's sugar cane production is more cost-effective than sugar as a feedstock for ethanol in the United States. A DOE study on world renewable fuels production potential lists sugar-based ethanol as the least expensive biofuel, but the price in the United States is higher than that cost because a tariff is levied on ethanol from Brazil.[191]

The United States has an ethanol tariff of $0.54 per gallon compared with $0.90 in Europe and Australia, $0.20 in Canada, and $0.27 in Central and South America.[192] Various subsidies help renewable fuels to compete with petroleum products, but are an issue of contention between the subsidizing countries and countries that want to enter the market.

Table A-1. Alternative Fuel Feedstock, Production Process, and Blending Option Comparison

Alternative	Renewable	Synthetic	Feedstock	Pre-processing	Product	Processing	Product	Production	Neat product	Blending	Final fuel	Fuel category	Comparable conventional fuel	Tactical fuel
Yes	Yes	No	Corn	Enzymatic Hydrolosis	Sugar	Fermenta-tion	Alcohols	Distillation	Ethanol/ Butanol	0%-85% blend w/ gasoline	E85, biobutanol	NTV Fuel	~Gasoline	No
Yes	Yes	No	Other Food Crops (Sugars)	Enzymatic Hydrolosis	Sugar	Fermenta-tion	Alcohols	Distillation	Ethanol/ Butanol	0%-85% blend w/ gasoline	E85, biobutanol	NTV Fuel	~Gasoline	No
Yes	Yes	No	Agricultural/ Forestry Residues	Enzymatic Hydrolosis	Sugar	Fermenta-tion	Alcohols	Distillation	Ethanol/ Methanol/ Butanol	0%-85% blend w/ gasoline	E85, M85, biobutanol	NTV Fuel	~Gasoline	No
Yes	Yes	No	Energy Crops (switchgrass, camelina, jatropa, etc)	Enzymatic Hydrolosis	Sugar	Fermenta-tion	Alcohols	Distillation	Ethanol/ Methanol/ Butanol	0%-85% blend w/ gasoline	E85, M85, biobutanol	NTV Fuel	~Gasoline	No
Yes	Yes	No	Urban wood/ mill waste	Enzymatic Hydrolosis	Sugar	Fermenta-tion	Alcohols	Distillation	Ethanol/ Methanol/ Butanol	0%-85% blend w/ gasoline	E85, M85, biobutanol	NTV Fuel	~Gasoline	No
Yes	Yes	No	Algae	Enzymatic Hydrolosis	Sugar	Fermenta-tion	Alcohols	Distillation	Ethanol/ Methanol/ Butanol	0%-85% blend w/ gasoline	E85, M85, biobutanol	NTV Fuel	~Gasoline	No
Yes	Yes	No	Fats/ Grease	N/ A	N/ A	N/ A	N/ A	Transeste-rfication	Biodiesel (FAME)	Blends of 5%-100% biodiesel w/ diesel	B5, B20, B100	NTV Fuel	Diesel	No
Yes	Yes	No	Food Oil Crops	Oil Extraction	Oils (Trigly-cerides)	N/ A	N/ A	Transeste-rfication	Biodiesel (FAME)	Blends of 5%-100% biodiesel w/ diesel	B5, B20, B100	NTV Fuel	Diesel	No

Alternative	Renewable	Synthetic	Feedstock	Pre-processing	Product	Processing	Product	Production	Neat product	Blending	Final fuel	Fuel category	Comparable conventional fuel	Tactical fuel
Yes	Yes	No	Energy Crops (switchgrass, camelina, jatropa, etc)	Oil Extraction	Oils (Trigly-cerides)	N/ A	N/ A	Transeste-rfication	Biodiesel (FAME)	Blends of 5%-100% biodiesel w/ diesel	B5, B20, B100	NTV Fuel	Diesel	No
Yes	Yes	No	Algae	Oil Extraction	Oils (Trigly-cerides)	N/ A	N/ A	Transeste-rfication	Biodiesel (FAME)	Blends of 5%-100% biodiesel w/ diesel	B5, B20, B100	NTV Fuel	Diesel	No
Yes	Yes	Yes	Fats/ Grease	N/ A	N/ A	Hydro-processing	Bio Crude	Fraction-ation	Bio-SPK	50/50 blend of Bio-SPK and diesel fuel	HRF-76	Renew-able (Marine) Diesel	Diesel	Yes
Yes	Yes	Yes	Food Oil Crops	Oil Extraction	Oils (Trigly-cerides)	Hydro-processing	Bio Crude	Fraction-ation	Bio-SPK	50/50 blend of Bio-SPK and diesel fuel	HRF-76	Renew-able (Marine) Diesel	Diesel	Yes
Yes	Yes	Yes	Energy Crops (switchgrass, camelina, jatropa, etc)	Oil Extraction	Oils (Trigly-cerides)	Hydro-processing	Bio Crude	Fraction-ation	Bio-SPK	50/50 blend of Bio-SPK and diesel fuel	HRF-76	Renew-able (Marine) Diesel	Diesel	Yes
Yes	Yes	Yes	Algae	Oil Extraction	Oils (Trigly-cerides)	Hydro-processing	Bio Crude	Fraction-ation	Bio-SPK	50/50 blend of Bio-SPK and diesel fuel	HRF-76	Renew-able (Marine) Diesel	Diesel	Yes
Yes	Yes	Yes	Fats/ Grease	N/ A	N/ A	Hydro-processing	Bio Crude	Fraction-ation	Bio-SPK	50/50 blend of Bio-SPK and Jet A (plus additives)	HRJ, HRJ-5, & HRJ-8	Hydro-treated Renew-able Jet (HRJ) Fuel	Jet-A, JP-5 & JP-8	Yes

Table A-1. (Continued)

Alternative	Renewable	Synthetic	Feedstock	Pre-processing	Product	Processing	Product	Production	Neat product	Blending	Final fuel	Fuel category	Comparable conventional fuel	Tactical fuel
Yes	Yes	Yes	Food Oil Crops	Oil Extraction	Oils (Trigly-cerides)	Hydro-processing	Bio Crude	Fraction-ation	Bio-SPK	50/50 blend of Bio-SPK and Jet A (plus additives)	HRJ, HRJ-5, & HRJ-8	Hydro-treated Renew-able Jet (HRJ) Fuel	Jet-A, JP-5 & JP-8	Yes
Yes	Yes	Yes	Energy Crops (switchgrass, camelina, jatropa, etc)	Oil Extraction	Oils (Trigly-cerides)	Hydro-processing	Bio Crude	Fraction-ation	Bio-SPK	50/50 blend of Bio-SPK and Jet A (plus additives)	HRJ, HRJ-5, & HRJ-8	Hydro-treated Renew-able Jet (HRJ) Fuel	Jet-A, JP-5 & JP-8	Yes
Yes	Yes	Yes	Algae	Oil Extraction	Oils (Trigly-cerides)	Hydro-processing	Bio Crude	Fraction-ation	Bio-SPK	50/50 blend of Bio-SPK and Jet A (plus additives)	HRJ, HRJ-5, & HRJ-8	Hydro-treated Renew-able Jet (HRJ) Fuel	Jet-A, JP-5 & JP-8	Yes
Yes	Yes	Yes	Agricultural/ Forestry Residues	Pyrolysis	Pyrolysis Oil	Hydro-processing	Bio Crude	Fraction-ation	Pyrolysis Renewable Jet (PRJ)	50/50 blend of PRJ and Jet A (plus additives)	SynJet	Cellu-losic Jet Fuel	Jet-A, JP-5 & JP-8	Yes
Yes	Yes	Yes	Energy Crops (switchgrass, camelina, jatropa, etc)	Pyrolysis	Pyrolysis Oil	Hydro-processing	Bio Crude	Fraction-ation	Pyrolysis Renewable Jet (PRJ)	50/50 blend of PRJ and Jet A (plus additives)	SynJet	Cellu-losic Jet Fuel	Jet-A, JP-5 & JP-8	Yes

Alternative	Renewable	Synthetic	Feedstock	Pre-processing	Product	Processing	Product	Production	Neat product	Blending	Final fuel	Fuel category	Comparable conventional fuel	Tactical fuel
Yes	Yes	Yes	Urban wood/ mill waste	Pyrolysis	Pyrolysis Oil	Hydro-processing	Bio Crude	Fraction-ation	Pyrolysis Renewable Jet (PRJ)	50/50 blend of PRJ and Jet A (plus additives)	PRJ	Cellulosic Jet Fuel	Jet-A, JP-5 & JP-8	Yes
Yes	Yes	Yes	Other Food Crops (Sugars)	Alcohol Oligomerization	Olefins	Polymeriz-ation w/ Hydroprocessing	Bio Crude	Fraction-ation	Catalytic Renewable Jet (CRJ)	50/50 blend of CRJ and Jet A (plus additives)	CRJ	SynJet	Jet-A, JP-5 & JP-8	Yes
Yes	Yes	Yes	Agricultural/ Forestry Residues	Gasification	Syngas	Fischer-Tropsch w/ Hydro-processing	SynCrude	Fraction-ation	Synthetic/ Iso-Paraffinic Diesels (IPD)	50/50 blend of IPD and diesel	FT-F-76	Cellulosic (Marine) Diesel	Diesel	Yes
Yes	Yes	Yes	Energy Crops (switchgrass, camelina, jatropa, etc)	Gasification	Syngas	Fischer-Tropsch w/ Hydro-processing	SynCrude	Fraction-ation	Synthetic/ Iso-Paraffinic Diesels (IPD)	50/50 blend of IPD and diesel	FT-F-76	Cellulosic (Marine) Diesel	Diesel	Yes
Yes	Yes	Yes	Urban wood/ mill waste	Gasification	Syngas	Fischer-Tropsch w/ Hydro-processing	SynCrude	Fraction-ation	Synthetic/ Iso-Paraffinic Diesels (IPD)	50/50 blend of IPD and diesel	FT-F-76	Cellulosic (Marine) Diesel	Diesel	Yes
Yes	Yes	Yes	Agricultural/ Forestry Residues	Gasification	Syngas	Fischer-Tropsch w/ Hydro-processing	SynCrude	Fraction-ation	Synthetic Paraffinic Kerosene (SPK)/ Iso-Paraffinic Kerosenes (IPK)	50/50 blend of SPK/ IPK and Jet A (plus additives)	FT-SynJet	Cellulosic Jet Fuel	Jet-A, JP-5 & JP-8	Yes

Table A-1. (Continued)

Alternative	Renewable	Synthetic	Feedstock	Pre-processing	Product	Processing	Product	Production	Neat product	Blending	Final fuel	Fuel category	Comparable conventional fuel	Tactical fuel
Yes	Yes	Yes	Energy Crops (switchgrass, camelina, jatropa, etc)	Gasification	Syngas	Fischer-Tropsch w/ Hydro-processing	SynCrude	Fraction-ation	Synthetic Paraffinic Kerosene (SPK)/ Iso-Paraffinic Kerosenes (IPK)	50/50 blend of SPK/ IPK and Jet A (plus additives)	FT-SynJet	Cellulosic Jet Fuel	Jet-A, JP-5 & JP-8	Yes
Yes	Yes	Yes	Urban wood/ mill waste	Gasification	Syngas	Fischer-Tropsch w/ Hydro-processing	SynCrude	Fraction-ation	Synthetic Paraffinic Kerosene (SPK)/ Iso-Paraffinic Kerosenes (IPK)	50/50 blend of SPK/ IPK and Jet A (plus additives)	FT-SynJet	Cellulosic Jet Fuel	Jet-A, JP-5 & JP-8	Yes
Yes	No	Yes	Coal	Gasification	Syngas	Fischer-Tropsch w/ Hydro-processing	SynCrude	Fraction-ation	Synthetic/ Iso-Paraffinic Diesels (IPD)	50/50 blend of IPD and diesel	FT-F-76	Synthetic (Marine) Diesel	FT-Diesel	Yes
Yes	No	Yes	Natural Gas	Gasification	Syngas	Fischer-Tropsch w/ Hydro-processing	SynCrude	Fraction-ation	Synthetic/ Iso-Paraffinic Diesels (IPD)	50/50 blend of IPD and diesel	FT-F-76	Synthetic (Marine) Diesel	FT-Diesel	Yes
Yes	No	Yes	Coal	Gasification	Syngas	Fischer-Tropsch w/ Hydro-processing	SynCrude	Fraction-ation	Synthetic Paraffinic Kerosene (SPK)/ Iso-	50/50 blend of SPK/ IPK and Jet A (plus	FT-SynJet, FT-SPK	Synthetic Jet Fuel	Jet-A, JP-5 & JP-8	Yes

Alternative	Renewable	Synthetic	Feedstock	Pre-processing	Product	Processing	Product	Production	Neat product	Blending	Final fuel	Fuel category	Comparable conventional fuel	Tactical fuel
									Paraffinic Kerosenes (IPK)	additives)				
Yes	No	Yes	Natural Gas	Gasification	Syngas	Fischer-Tropsch w/ Hydro-processing	SynCrude	Fraction-ation	Synthetic Paraffinic Kerosene (SPK)/ Iso-Paraffinic Kerosenes (IPK)	50/50 blend of SPK/ IPK and Jet A (plus additives)	FT-SynJet, FT-SPK	Synthetic Jet Fuel	Jet-A, JP-5 & JP-8	Yes

International oil market developments likewise impact the use of renewable fuels as the price of oil rises and falls. Policies to stabilize biofuel prices vis a vis petroleum prices have limited impacts on demand for biofuel in the United States.

GHG regulation in other countries also affects international markets for renewable fuels. For example, there is a proposal in the EU to include aviation fuels as part of the EU ETS. The EU ETS requires countries to limit GHGs but enables the trading of allowances. This proposal would assign and auction allowances to airline operators at a 3 percent reduction from historical emissions in 2012 and a percent reduction in 2013.[193] However, it excludes military, humanitarian, and government flights among several other categories.

Consumer Demand for Renewable Fuels

Consumer demand for renewable fuels is limited in part by the availability of infrastructure to supply these fuels. Ethanol in the form of E85 is only available at 1.2 percent of all fuel stations in the United States, mostly close to production facilities in the Midwest. B20 is available at 0.4 percent of fuel stations, but they are dispersed throughout the United States. E85 is also limited by the availability of alternative fuel vehicles.

The amount of renewable fuel that can be blended with conventional fuels is limited. For example, ASTM D975 allows conventional diesel to contain up to 5 percent biodiesel (B5).[194] Biodiesel consumption is complicated by warranty issues, cost, and concerns about engine and filter performance. Existing legislation allows ethanol blends up to 10 percent (E10) as fuel additive to gasoline, but this market will be saturated by 2015. However, recent regulatory developments suggest that this threshold could increase to E15 prior to the 2015 blending wall.

APPENDIX B. ETHANOL DETAILED MARKET PROJECTIONS

In this appendix, we detail our assessment of the current and future commercial availability of ethanol. Corn-based ethanol currently represents the largest renewable fuel in the United States: more than 10 billion gallons were produced in 2009. Most of this ethanol (99.5 percent) is sold as "gasohol" or "E10" (typically a 90 percent gasoline and 10 percent ethanol blend),[195] and the remainder is sold as E85 (typically a blend of 85 percent or more of denatured ethanol with gasoline). Other ethanol feedstocks represent only a small portion of production in the United States.

The market for ethanol is forecast to increase dramatically over the next 10 years, driven by the RFS2 mandate and the emergence of cellulosic ethanol. The projected ethanol production quantities will likely exceed both the E10 and E15 markets, necessitating substantial increased blending and sales of E85. The major hurdle to using E85 (over this study's time frame) is commercial availability, which is limited and highly regionalized. In this appendix, we discuss the projected commercial availability of ethanol blended as E10, E15, and E85 from 2010 through 2020.

To forecast their commercial availability, we consider each supply chain stage. The supply chain includes feedstock production and distribution, fuel production, transportation of fuel to bulk facilities for blending with conventional fuels, and finally, transportation to a retail facility for final sale to consumers. Market conditions, costs, and bottlenecks at each of these stages of the supply chain determine the availability of biofuels.

Summary

We forecast that national ethanol retail sales will increase more than 73 percent to 18,660 million gallons by 2020 (Table B-1). In the short term (through 2013), the primary factor limiting ethanol availability is availability of corn feedstock. As production capacity for cellulosic ethanol comes online after 2013, cellulosic production capacity will be the bottleneck in ethanol production.

The federal RFS and RFS2 programs have bolstered the demand for ethanol, and its chemical properties make it a relatively good gasoline additive as an oxygenate and octane enhancer. We forecast that almost all (more than 99.3 percent) ethanol will be blended as E10 or E15 until the fuel additive market is saturated in 2015. In 2016 and onward, we expect the E85 market to grow significantly as almost all increases in ethanol production are blended as E85. Over the next 5 years, commercial fueling infrastructure will develop to support these projected E85 sales.

Supply Chain Overview

Ethanol, or ethyl alcohol, is produced through the fermentation and distillation of simple sugars. Ethanol can be made from a wide array of biological feedstocks that contain either substantial amounts of sugar or materials that can be converted into sugar (such as starch or cellulose). In the United States, most ethanol is currently produced from corn: the starch in corn is readily converted into sugar. Other potential feedstocks include grasses (cellulose) and sugars, but their production volumes are limited due to cost and feedstock availability. Cellulosic ethanol is expected to emerge as renewable fuel within the next few years due to both mandates and improved commercial viability.

As of January 2010, 189 ethanol refineries were operating in the United States, at a total annual production capacity of 11.9 billion gallons (4.9 billion gallons in 2006).[196] An additional 16 biorefineries are being built or expanded that will add 1.4 billion gallons of new capacity.[197] Coupled with idle capacity of 1.2 billion gallons, the total ethanol annual production capacity is expected to increase to 14.5 billion gallons within the next few years. Almost all of these refineries produce ethanol from corn, so they are concentrated near the feedstock source, the "Corn Belt."

Denatured ethanol is transported from the refinery to either a bulk terminal or a redistribution bulk terminal (Figure B-1). Most ethanol is transported to terminals on the freight rail system, and the remainder is transported via tanker truck or barge.[198] From the terminals, ethanol is transported (typically after blending with gasoline) to retail locations by tanker truck for sale to end users. Most ethanol (99.5 percent) is sold as E10,[199] and the remainder is sold as E85.

Table B-1. Biodiesel and Renewable and Cellulosic Diesel Forecasts for Supply Chain Stages, 2009–20 (million gallons)

Stage	Type	2009[a]	2010	2011	2012	2013	2014	2015	2016	2017	2018	2019	2020
Feedstock availability	*Conventional Ethanol*												
	Corn[b]	10,606	11,810	12,430	12,910	13,395	13,870	14,140	14,400	14,600	14,800	15,000	15,075
	Others[c]	246	248	278	297	307	315	329	341	346	350	352	355
	Total	10,852	12,058	12,708	13,207	13,702	14,185	14,469	14,741	14,946	15,150	15,352	15,430
	Cellulosic Ethanol												
	Cellulosed	33.0 to 99.4 Billion Gallons											
	Total	10,852	12,058	12,708	13,207	13,702	14,185	14,469	14,741	14,946	15,150	15,352	15,430
Production capacity	*Conventional Ethanol*												
	Corn[e]	11,649	12,251	12,853	13,455	14,057	14,293	14,529	14,764	15,000	>15,000	>15,000	>15,000
	Others[e]	225	248	278	297	307	315	329	341	346	350	352	355
	Total	11,874	12,499	13,131	13,752	14,364	14,608	14,858	15,105	15,346	>15,350	>15,352	>15,355
	Cellulosic Ethanol												
	Cellulos[e]	3	3	8	167	344	600	1,022	1,456	1,878	2,389	2,900	3,589
	Total		12,502	13,139	13,919	14,708	15,208	15,880	16,561	17,224	17,739	18,252	18,944
Projected production	*Conventional Ethanol*												
	Corn	10,606	11,810	12,430	12,910	13,395	13,870	14,140	14,400	14,600	14,800	15,000	15,075
	Others	150	248	278	297	307	315	329	341	346	350	352	355
	Total	10,756	12,058	12,708	13,207	13,702	14,185	14,469	14,741	14,946	15,150	15,352	15,430
	Cellulosic Ethanol												
	Cellulose[f]	2	5	7.3	150	310	540	920	1,310	1,690	2,150	2,610	3,230
	Total		12,063	12,715	13,358	14,012	14,727	15,387	16,057	16,639	17,301	17,962	18,660
Retail sales	E85[d]	54	58	61	69	76	81	101	1,139	1,486	1,844	2,431	3,229
	E10 or E15	10,702	12,005	12,654	13,289	13,936	14,646	15,286	14,918	15,153	15,457	15,531	15,431
	Total[d,e,f]	10,756	12,063	12,715	13,358	14,012	14,727	15,387	16,057	16,639	17,301	17,962	18,660

[a] 2009 values represent actual, not projected, data.

[b] U.S. Department of Agriculture (USDA), *USDA Agricultural Projections to 2019,* OCE-2010-1, February 2010.
[c] Food and Agricultural Policy Research Institute (FAPRI) at The University of Missouri, *U.S. and World Agricultural Outlook,* 2010.
[d] EIA, *Annual Energy Outlook 2010*, DOE/EIA-0383(2010), May 11, 2010.
[e] Renewable Fuels Association, *2010 Ethanol Industry Outlook,* February 2010, www.ethanolrfa.org/ pages/annual-industry-outlook.
[f] EPA, *Renewable Fuel Standard Program (RFS2) Regulatory Impact Analysis*, EPA-420-R-10-006, February 2010.

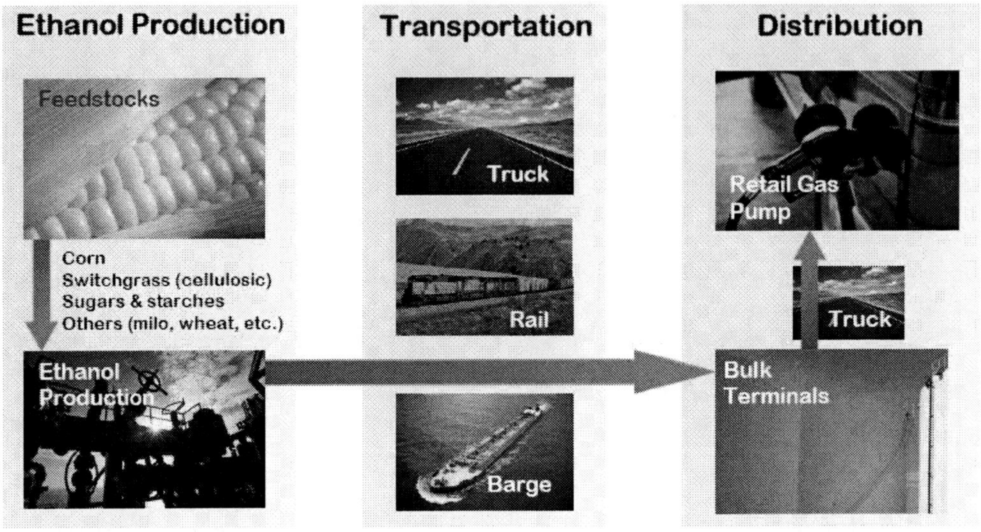

Figure B-1. Ethanol Supply Chain.

Ethanol Feedstocks

Summary

The forecast for feedstocks for ethanol production through 2020 is as follows:

- *Corn will continue as the feedstock for roughly 98 percent of firstgeneration ethanol produced.* Corn will remain the preferred feedstock for ethanol production due to its availability, commodity cost, and economics for ethanol production. Corn is the only first-generation feedstock that enables ethanol to compete with the cost of petroleum at the scale necessary to meet demand (billions of gallons annually).
- *Corn ethanol production will reach the limit of corn availability (15 billion gallons) in 2019.* The theoretical limit of corn available for conversion into ethanol is 15 billion gallons. We project corn ethanol production to meet this limit in 2019, and remain near that level over the long term.
- *Sugar will not significantly contribute to the ethanol market.* Sugar is grown in only four U.S. states and Puerto Rico, so it is not available on the scale necessary to produce significant quantities of ethanol. The availability of sugar as a feedstock for ethanol production is further limited by its importance in the food market—diverting its use to ethanol would compete with its use as a food product. In addition, although sugar cane has a higher yield than corn for ethanol production, its domestic commodity cost leads to a much higher ethanol cost per gallon than corn.
- *The cellulosic pathway will achieve commercial viability around 2012 and support most ethanol production growth afterward.* Second-generation ethanol production from cellulosic biomass generates less GHG emissions, has an abundance of available feedstock, and provides production costs far less than corn-based ethanol. EPA and DOE project commercial demonstration of a viable pathway by 2012 and commercially viable ethanol production from cellulosic biomass beyond that date.

Comparison of Current Feedstock Yields and Costs

As shown in Figure B-2, of the first-generation ethanol feedstocks, sugar cane and sugar beets have the highest yield in gallons per acre, 37 percent higher than corn. However, due to corn's lower commodity cost and byproducts—production of ethanol from U.S.-harvested sugar cane costs $2.77 per gallon compared with corn grain at $1.46 per gallon of ethanol at current commodity prices.[200] Figure B-3 shows ethanol production costs for each of the primary feedstocks.

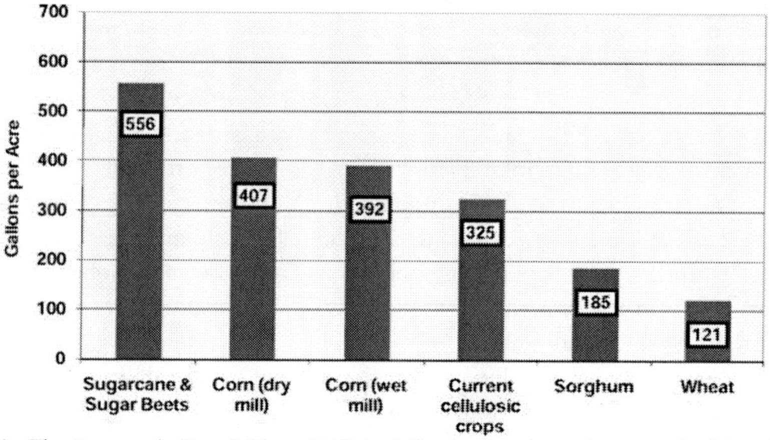

Source: USDA, The Economic Feasibility of Ethanol Production from Sugar in the United States, July 2006.

Figure B-2. Ethanol Production Yields for Primary Feedstocks (gallons per acre).

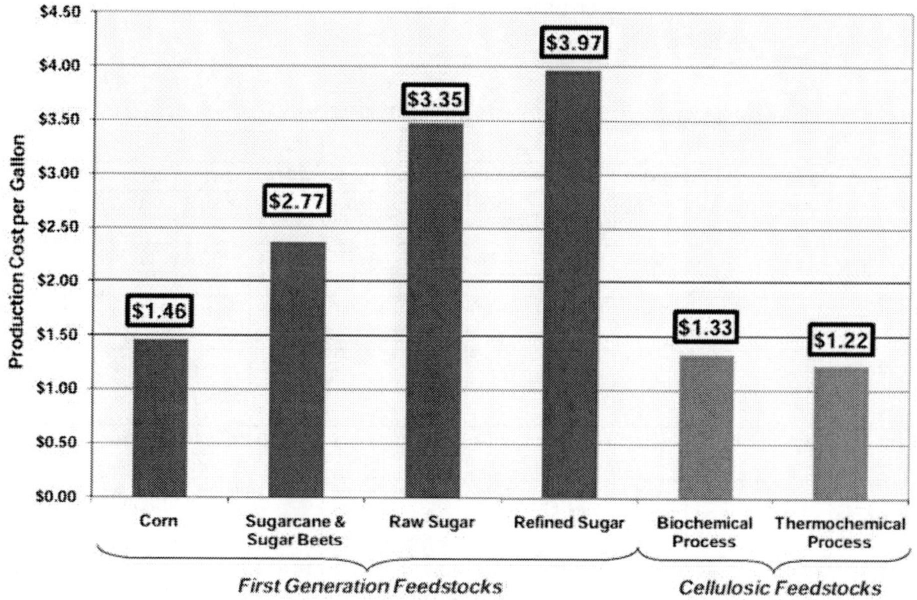

Source: USDA, *The Economic Feasibility of Ethanol Production from Sugar in the United States*, July 2006.

Figure B-3. Ethanol Production Costs for Primary Feedstocks ($ per gallon).

First-Generation Ethanol

Almost all of the ethanol produced today is considered a first-generation biofuel, or a biofuel generated from crops or animal fats that can also be used as food and feed. The primary potential first-generation feedstocks for ethanol production in the United States through 2020 are as follows:

- *Corn.* Although estimates vary, roughly 98 percent of the ethanol produced in the United States comes from corn. The production of ethanol from corn requires the use of enzymes to break down the corn starch into sugars, which are then fermented into ethanol.
- *Sugars and starches (first-generation biofuel feedstock).* Sugar-based feedstocks include sugar beets, sugar cane, potato starches, and sweet sorghum. Ethanol production using sugar-based feedstocks is more efficient because the initial enzymatic step required for corn processing is unnecessary. However, only one refinery in the United States produces ethanol from sugar, and its annual capacity is only 1.5 million gallons. Growing conditions (sugar beets must be rotated with other crops) and government tariffs are primarily responsible for limiting sugar as a feedstock in the United States.
- *Other Feedstocks.* Other feedstocks for ethanol production include milo, wheat, sorghum, barley, brewery waste products, and cheese whey. However, feedstock availability and policy preferences will limit their use over the time frame of this study.

Corn

The USDA reported that 86.5 million acres of corn were planted in 2009, yielding 12,086 million bushels (or 153.9 bushels per harvested acre).[201] As shown in Figure B-4, the majority of this acreage is located in the Midwest.

Table B-2. Forecast of Potential Ethanol Production from Corn, 2009–20

	2009	2010	2011	2012	2013	2014	2015	2016	2017	2018	2019	2020
Million acres planted	86.5	86.4	88	90	89.5	89.5	89.5	89.5	89.5	89.5	89.5	89
Yield (bushels per acre)	153.9	162.9	160.4	162.4	164.4	166.4	168.4	170.4	172.4	174.4	176.4	178.4
Corn produced[a]	12,086	12,921	12,960	13,445	13,530	13,695	13,860	14,025	14,190	14,355	14,520	14,595
Corn available for ethanol use[a]	3,788	4,200	4,400	4,550	4,700	4,825	4,875	4,925	4,950	4,975	5,000	5,025
Corn available for other use[a]	8,424	8,721	8,560	8,895	8,830	8,870	8,985	9,100	9,240	9,380	9,520	9,570
Ethanol conversion rate[b]	2.8	2.81	2.83	2.84	2.85	2.88	2.9	2.93	2.95	2.98	3	3
Potential ethanol production[a]	10,606	11,810	12,430	12,910	13,395	13,870	14,140	14,400	14,600	14,800	15,000	15,075

[a] Millions of bushels.

[b] Gallons per bushel.

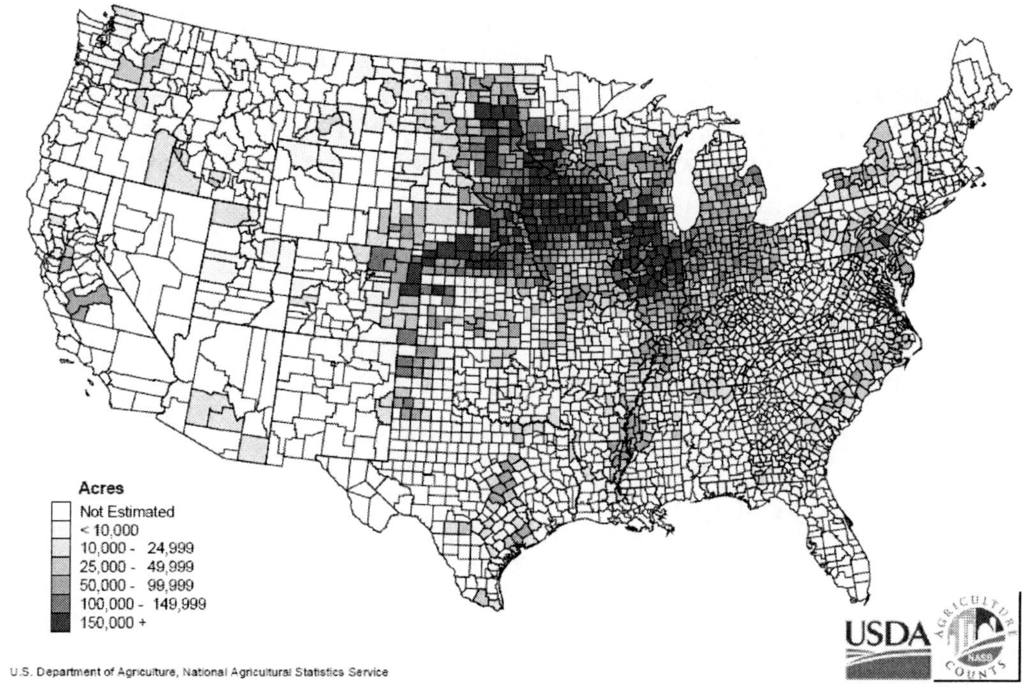

Source: USDA, National Agricultural Statistics Service, July 2010.

Figure B-4. Planted Acres of Corn for Grain by County (2008).

Annual crop yields have increased over the last 10 years from 137.1 bushels per acre in 2000 to 162.9 in 2010.[202] USDA projects that advances in technology will further increase yields to 178.4 bushels per acre in 2020.[203]

The USDA projects that the corn available annually for ethanol production will rise sharply from 3,677 to 5,025 million bushels from 2009 to 2020. This rise reflects projected corn production above a "baseline" demand for non-ethanol uses of corn between 8.4 and 9.6 billion bushels from 2009 to 2020. From this growth in available corn—coupled with projected slight increases in the conversion rate of ethanol from corn—we project that the potential ethanol that can be produced from corn will increase 42 percent, from 10,606 to 15,075 million gallons between 2009 and 2020 (Table B-2 and Figure B-5).

Less than half of the projected growth in corn production will be used to meet increased demand for ethanol production. Corn supply growth will result from both increased yields and acreage planted. Since we expect corn supply to exceed demand for ethanol use over the next 10 years, corn prices will likely remain relatively stable over this time period.

Corn Prices

One of the key factors keeping ethanol cost-competitive with gasoline is the price of corn. According to DOE's AEO, corn feedstock constitutes 57 percent of ethanol production costs.

The cost of corn depends on the supply and demand dynamics of the corn market. Corn is the largest crop (by acreage) in the United States. A staple in the food and agricultural feed industries, it is used for everything from tortillas to beverage sweeteners to animal feed. As the demand for corn from any of the various market sectors that depend on it increases, annual reserve stocks decrease and prices increase. These price increases affect the end prices of all products that use corn, including ethanol.

USDA forecasts that corn production will expand to meet ethanol requirements over the next 10 years, and corn prices will stabilize in 2013 near $3.70 per bushel (Figure B-6).[204]

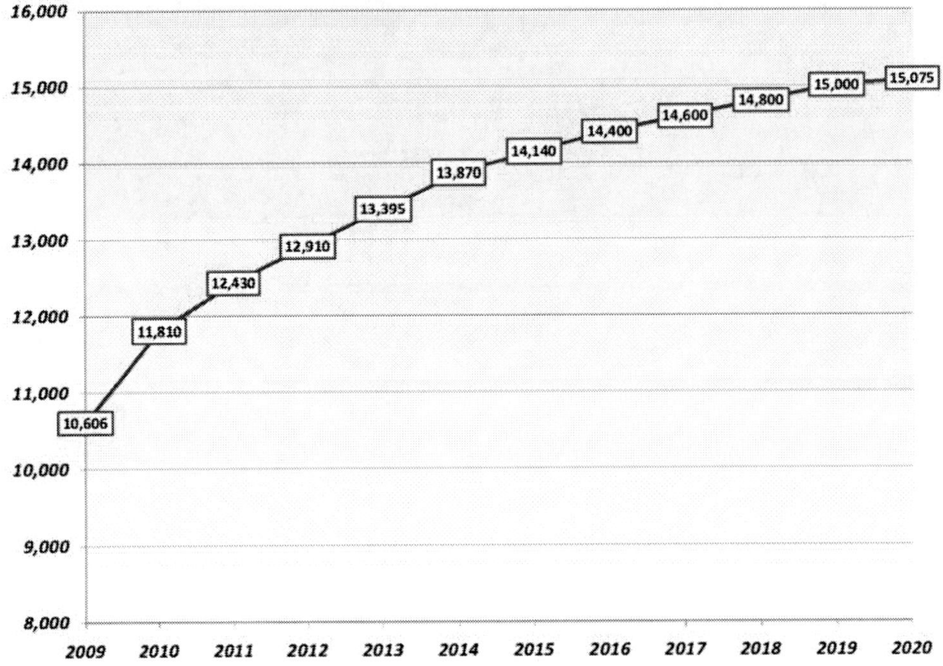

Figure B-5. Forecast of Potential Ethanol Production from Corn, 2009–20 (million gallons).

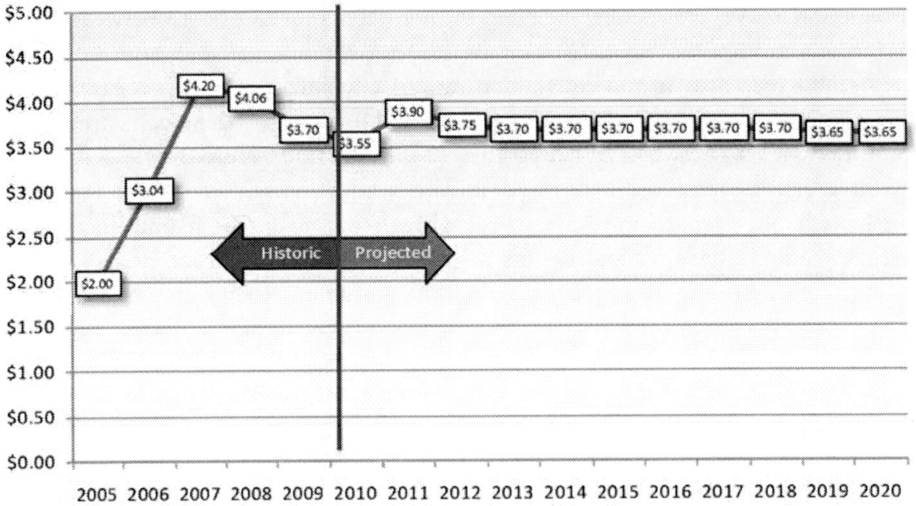

Sources: USDA National Agricultural Statistics Service, Corn National Statistics; USDA, *Agricultural Projections to 2019*, 2010.

Figure B-6. Historic and Forecast Corn Prices, 2005–20.

If corn prices behave as forecast, ethanol will continue to be cost-competitive with gasoline. Given current tax credits for ethanol production, Aventine Renewable Energy estimates that for ethanol to compete with gasoline at $30 per barrel of crude oil, corn prices must be less than $3.00 per bushel. At $40 per barrel of crude oil, ethanol is cost-competitive at corn prices less than $4.00.[205] If crude continues to cost well over $40 per barrel, a $4 per bushel cost of corn is probably sustainable.

Without tax credits for ethanol production, the landscape would change. Goldman Sachs estimates that if tax credits were taken away, new ethanol refinery builds would only be justified at crude oil prices above $63 per barrel (at projected longterm corn prices of $3.70 per bushel). If corn prices rise to $4.00 per bushel, ethanol production would expand only at crude oil prices above $65 per barrel. At $3.25 per bushel, such expansion would occur only at $55 per barrel and at $2.50 per bushel, expansion would occur at a crude oil cost above $45 per barrel.[206]

Corn Availability for Food

Some argue that the increased percentage of corn used for ethanol raises a "foodversus-fuel" concern. However, the use of corn as a feedstock for ethanol has had relatively little effect on the availability of corn for food. We estimate that due to increased corn harvest yields and ethanol conversion rates, the supply of corn for non-ethanol uses will increase by 14 percent from 2009 through 2020. Impacts on animal feed are also expected to be minimal due to ethanol production byproducts, DDGS. DDGS have been introduced to the market relatively recently, and advances in related technologies and infrastructure may allow them to displace up to 1 billion bushels of corn as animal feed annually.[207]

USDA's June 2010 crop report suggests that increased production of ethanol is not having significant effects on the acreage of land dedicated to crops. In 2010, land use dedicated to crops decreased six million acres compared to 2008, and over eight million compared to the average in the 1990s. Despite the acreage decrease, a record crop is expected in 2010, primarily due to increased yields.

Recent growth in the use of ethanol has only generated limited impacts on the retail price of food. The CBO found that expanded production of ethanol between April 2007 and April 2008 resulted in an increase of 0.5 to 0.8 percentage points of the 5.1 percent increase in food prices measured by CPI.[208] Higher energy costs had a larger effect on food prices than ethanol production; other factors included growing demand for meat, depreciation of the U.S. dollar, and harvest concerns.

The EPA, in the Final Rule on the Renewable Fuel Standard Program (March 2010), found that the RFS2 will have an impact of $10 more per capita on food compared to the AEO 2007 reference case.[209]

In 2006, roughly 20 percent of the corn crop was used for ethanol. In 2010, due to rapid expansion in ethanol production, we estimate that approximately 32 percent of the crop will be used for ethanol. Our projects for corn-based ethanol production suggest that by 2020, this percentage will rise only by 2 percentage points, much less than the previous increases.

Sugar Cane and Sugar Beets

Ethanol production from sugar cane and sugar beets is limited due its high cost and limited availability relative to corn. Currently, there are no refineries in the United States is producing ethanol from sugars, and no new sugar refineries are planned.

As shown in Figure B-7, sugar cane is grown in Texas, Louisiana, Hawaii, and Florida. Sugar beets are grown in Minnesota, North Dakota, and several western states (see Figure B-8). In 2009, U.S. sugar cane production was 28.4 million tons and U.S. sugar beet production 29.5 million tons.[210] At this production level, total potential ethanol production using the entire harvest of domestic sugar cane and domestic sugar beets is only 554 million gallons and 732 million gallons, respectively. Combined, this theoretic production represents roughly 12 percent of the ethanol that was produced from corn in 2009.

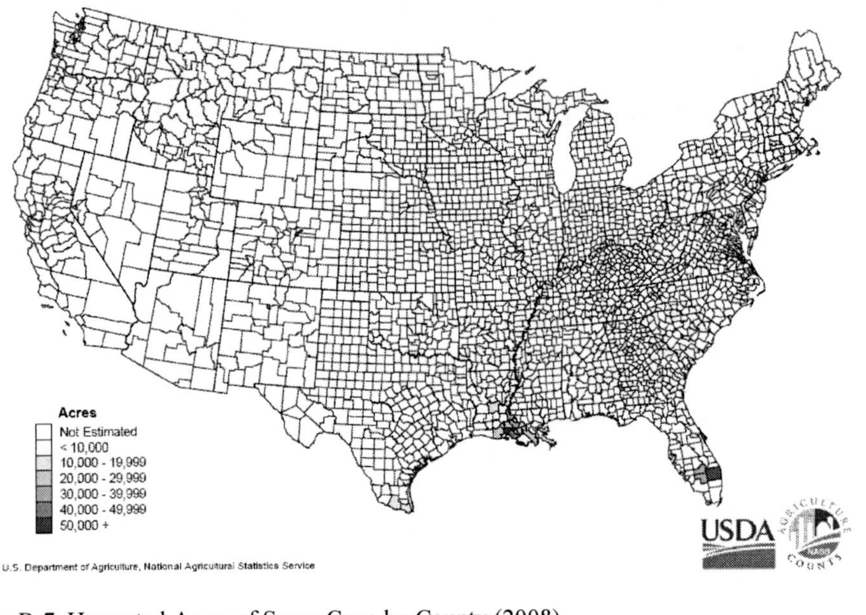

Figure B-7. Harvested Acres of Sugar Cane by County (2008).

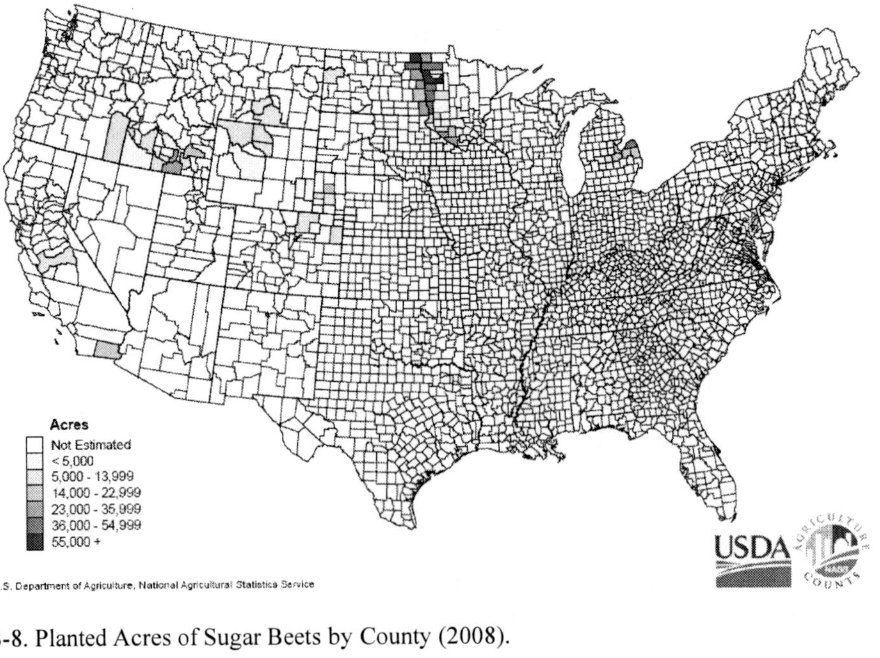

Figure B-8. Planted Acres of Sugar Beets by County (2008).

Even if availability were not an issue, high feedstock and processing costs limit the current commercially viability of ethanol produced from sugar. At today's domestic sugar prices, the cost of converting sugar cane and sugar beets to ethanol is roughly $2.60 per gallon.[211] Feedstock costs represent between 65 and 70 percent of total production costs. With current (May 27, 2010) ethanol prices around $1.62 per gallon, producing ethanol from sugar is unprofitable.

It is important to note that ethanol from sugar cane qualifies to count as an advanced biofuel for RFS2. However, in order to remain price competitive, feedstocks would have to be imported, and U.S. tariffs lowered. EPA estimates that 1.03 billion gallons of sugar cane ethanol will be imported to the United States in 2022 under RFS2.[212]

Second-Generation (Cellulosic) Ethanol

Second-generation biofuels include biofuels produced from cellulose hemicellulose or lignin. Cellulosic feedstocks include corn stover, timber wastes, and dedicated energy crops such as switchgrass. Ethanol is produced from these feedstocks by isolating sugar molecules in the plant cell walls and converting them into ethanol. Technology enabling the production of ethanol from cellulosic materials is still in its early stages (several pilot plants are in operation), but the market is expected to grow dramatically over the next 10 years, driven primary through the RFS2.

Cellulosic Feedstocks

Cellulosic biomass feedstocks offer the greatest potential for producing quantities of ethanol necessary to significantly displace petroleum. Some cellulosic crops have a higher yield per acre planted than corn—800 gallons compared with 416. Since these feedstocks can be grown on marginal lands with low energy, water, and fertilizer requirements, cellulosic biomass may be produced at much higher quantities and lower costs than corn.

Cellulosic biomass is composed of a wide variety of forest and agricultural residues, industrial processing residues, and municipal solid and urban wood residues.[213] Table B-3 outlines the four biomass resources categories, and the primary cellulosic biomass resources within each category.

Table B-3. Types of Cellulosic Biomass Resources

Agricultural resources	Energy crops	Forest residues	Urban wood waste and mill residues
Crop residues from major crops (corn stover, small grain straw, and others) Cellulosic components of separated food wastes and MSW	Perennial grasses (switchgrass, sorghum, miscanthus, and energycane) Perennial woody crops	Logging residues from conventional harvest operations and residues from forest management and land clearing operations Removal of excess biomass (fuel treatments) from timberlands and other forestlands Fuelwood extracted from forestlands	Wood processing mill residues (primary and secondary) Pulping liquors Urban wood residues (construction and demolition debris, tree trimmings, and packaging wastes) Perennial woody crops

Table B-4. Cellulosic Biomass Available for Conversion into Biofuel (million dry tons)

Biomass category	DOE/USDA billion-ton study	EIA 25x'25 study	UCS climate 2030 study
Agricultural residues	371–534	158	158
Energy crops	156–377	242	121
Forest residues	140	168	62
Urban wood waste and mill residues	63	29	27
Total available biomass	730–1,114	598	368

Table B-5. Potential Biofuel Production from Available Cellulosic Biomass (billion gallons)

Biomass category	Conversion rate[a] (gal/ton)	DOE/USDA billion-ton study	EIA 25x'25 study	UCS climate 2030 study
Agricultural residues	92.3	34.2–49.3	14.6	14.6
Energy crops	80	12.5–30.2	19.4	9.7
Forest residues	101.5	14.2	17.1	6.3
Urban wood waste and mill residues	90	5.7	2.6	2.4
Total potential production		66.6–99.4	53.7	33.0

[a] EPA, *Renewable Fuel Standard Program (RFS2) Regulatory Impact Analysis*, EPA-420-R-10-006, February 2010.

Estimates of the amount of biomass available for conversion into biofuel vary greatly. Table B-4 shows the biomass estimates from three studies: (1) the DOE and USDA Billion-Ton study,[214] the EIA 25x'25 study,[215] and the Union of Concerned Scientists (UCS) Climate 2030 study.[216] Table B-5 presents the theoretical maximum potential biofuel that can be produced from the available cellulosic biomass.

Forest Residues, Urban Wood Waste, and Mill Residues

Forest resources, urban wood waste, and mill residues available for cellulosic ethanol production are found in the wooded regions of the United States (see Figure B-9). These include the Pacific Northwest, the deep South, Canadian border regions in the Midwest, and the Northeast.

Agricultural Residues and Energy Crops

As shown in Figure B-10, agricultural residues and energy crops are primarily located in the Corn Belt, Lake States, Appalachian, and Delta, and Southeast regions. Corn stover production is dominant in the Corn Belt and Lake States regions, straw in the Northern Plains, Mountain States, and Pacific regions, and energy crops in the Delta, Appalachian, Corn Belt, and Southeast regions.

Feedstock Costs

"The economic competitiveness of cellulosic ethanol production is highly dependent on feedstock cost, which constitutes 35 to 50 percent of the total ethanol production cost, depending on various geographical factors and the types of systems used for harvesting, collecting, preprocessing, transporting, and handling the material."[217] Feedstock costs can be broken down into the following categories:

- *Harvest and Collection.* Each potential feedstock will be harvested slightly differently and will have different prices as a result. Some could be baled and stacked, others (such as woody sources) flown out by helicopter.
- *Storage and Queuing.* Biomass feedstocks, particularly agricultural ones, have a relatively narrow harvest window, despite being needed year-round at the biorefinery. As a result, they must be stored at intermediate facilities. Microbial losses at these facilities vary between 1 percent and 25 percent of the total crop and depend on the range of environmental factors, including moisture content, humidity, rainfall, and temperature. Woody feedstocks, which are typically 50 percent moisture following harvest, can be stored at this moisture content without degradation[218] while agricultural feedstocks over about 15 percent moisture content are at greater risk.
- *Preprocessing.* Cellulosic materials are often bulky and difficult to transport. For example, corn cobs are 2 to 4 times as expensive to transport as corn kernels because of additional bulk.[219] In order to facilitate transportation, storage, and processing, the materials need to be preprocessed, including grinding.
- *Transportation and Handling.* The cost of transporting and storing bulky cellulosic biomass material is not well understood, hence is often undervalued.[220] Feedstock moisture content can increase weight and volume without increasing cellulosic yield, thus increasing shipping costs for each unit of cellulosic material.

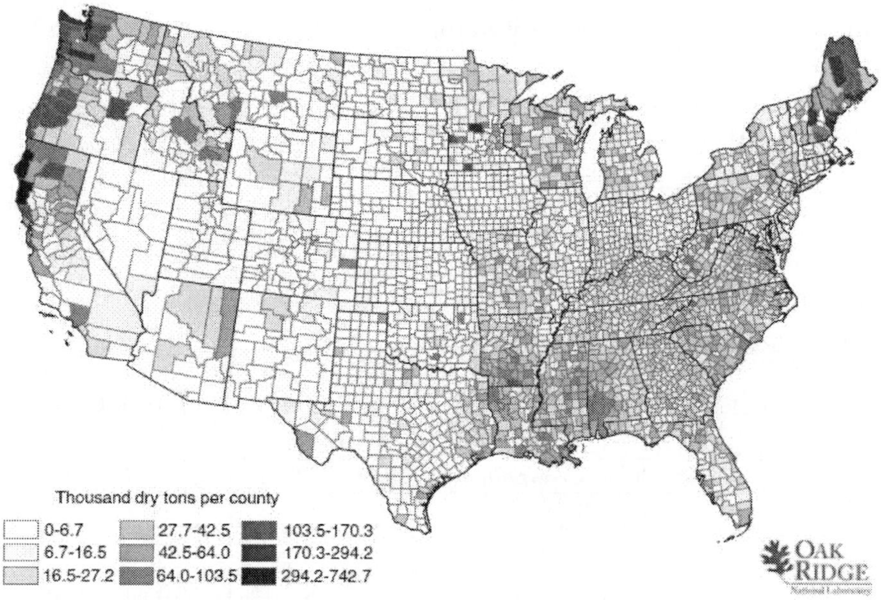

Source: Biomass Research & Development Board, "Increasing Feedstock Production for Biofuels," Feedstock Economic Analysis, 2008.

Figure B-9. United States Forestland Biomass Resources.

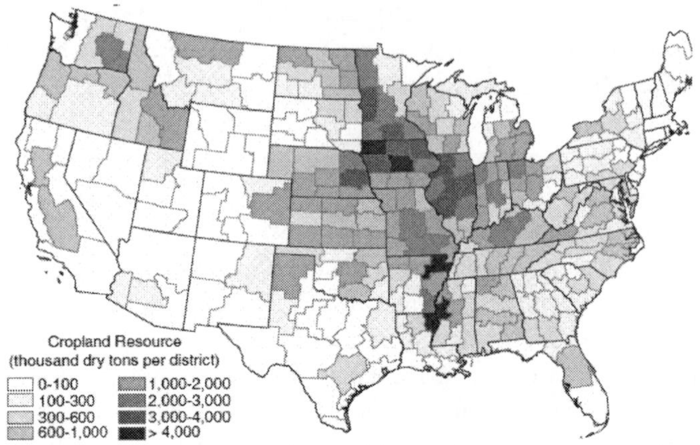

Source: Biomass Research & Development Board, "Increasing Feedstock Production for Biofuels," Feedstock Economic Analysis, 2008.

Figure B-10. United States Cropland Biomass Resources.

Cellulosic feedstocks are typically classified as wet or dry herbaceous, to address both storage degradation and transportation issues. Dry feedstocks, which have moisture contents less than 20 percent, include cereal straw and switchgrass. Wet feedstocks, which have moisture contents greater than 40 percent, include corn stover, sorghum stover, and forest residues.

Tables B-6 and B-7 provide estimates of current and projected feedstock costs for dry and wet feedstocks. The costs are broken out by each of the feedstock cost categories listed above.

Table B-6. Current and Projected Dry Herbaceous Cellulosic Feedstock Costs, 2009–17

Feedstock cost component	2009	2012	2017
Harvest and collection	$12.40	$12.15	$10.80
Storage and queuing	$6.75	$5.95	$5.30
Preprocessing	$12.35	$10.75	$8.00
Transportation and handling	$10.10	$6.15	$5.90
Total cost per dry ton	$41.60	$35.00	$30.00

Source: DOE, Office of the Biomass Program, Energy Efficiency and Renewable Energy, *Biomass Multi-Year Program Plan*, 2010, p. 3–21.

Table B-7. Current and Projected Wet Herbaceous Cellulosic Feedstock Costs, 2009–17

Feedstock cost component	2009	2012	2017
Harvest and collection	$20.70	$10.60	$10.60
Storage and queuing	$17.80	$11.10	$8.60
Preprocessing	$11.50	$8.70	$7.80
Transportation and handling	$16.10	$14.70	$14.70
Total cost per dry ton	$66.10	$45.10	$41.70

Source: DOE, Office of the Biomass Program, Energy Efficiency and Renewable Energy, *Biomass Multi-Year Program Plan*, 2010, p. 3–21.

Ethanol Production

First-Generation Ethanol

As of January 2010, 187 ethanol refineries (not including cellulosic ethanol refineries) were operating in the United States, representing a total production capacity of 11.9 bgpy (2009 production reached 10.6 billion gallons). Eleven ethanol refineries representing an additional production capacity of 1.2 bgpy are not currently operating. Ten new ethanol refineries are under construction and 5 are expanding, which will increase total production capacity to 14.4 bgpy over the next few years.[221] This increase in capacity will phase in over the next several years as construction projects are completed.

Currently, almost all ethanol refinery operating capacity uses corn as the primary feedstock—11,649 million gallons (or 98.1 percent) of the 11,877 million gallons of total capacity (includes corn fraction when combined with other feedstocks).[222] This will continue in the foreseeable future because almost all of the idle capacity (98.7 percent) and most (89 percent) of the planned expansion of production capacity in the next few years will utilize corn feedstocks. As shown in Figure B-11, after planned capacity comes online, ethanol production from corn (includes corn fraction when combined with other feedstocks) will constitute 12,508 million gallons (or 98.6 percent) of the 12,687 million gallons of total refining capacity.

As shown in Figure B-12, most of the current and planned refineries are located in the Corn Belt near the feedstock source. Of the currently operating ethanol plants, most are located in Iowa, Minnesota, Nebraska, and South Dakota, which produce almost half of the U.S. corn crop. The industry has been slowly expanding beyond the Corn Belt, as facilities recently came online in Oregon, New York, and Texas, and planned facilities will come online along the East coast, in Texas, and California.

Biorefineries produce corn-based ethanol using either dry or wet milling techniques (dry milling is the most common). Current dry milling techniques can produce 2.75 gallons of ethanol per bushel of corn (roughly 400 gallons/acre), while wet milling techniques yield 2.65 gallons per bushel (390 gallons/acre). The primary difference between the two processes is the pretreatment of the incoming grain—dry milling uses hammer mills to grind the grain into a starch-containing powder, while wet milling uses a liquid solution to separate the grain into a range of constituent parts, including starch. For either process, once the initial grain pretreatment occurs, enzymes convert the starch into sugars, which are then fermented into ethanol. The final step is to denature the ethanol by adding a small amount of gasoline to render it undrinkable.

Ethanol production from corn also generates byproducts that result in additional revenue for the refinery. In the dry milling process, roughly 6 pounds of DDGS, sold for use in animal feed, are produced per gallon of ethanol. The byproducts of the wet milling process include roughly 4.9 pounds of corn gluten feed, 0.9 pounds of corn gluten meal, and 0.6 pounds of corn oil per gallon of ethanol.[223]

Second-Generation Ethanol

As of January 2010, two cellulosic ethanol refineries were operating in the United States, representing a total production capacity of 3 mgpy. One new cellulosic ethanol refinery is under construction with a production capacity of 100 mgpy.

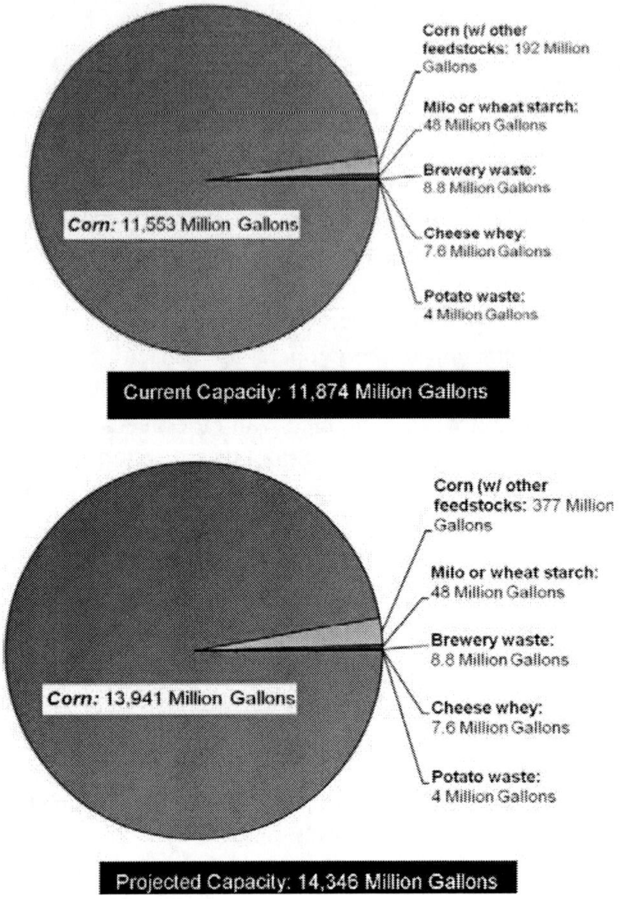

Figure B-11. Current and Planned Capacity of Non-Cellulosic Ethanol Refineries by Feedstock.

We project growth in cellulosic ethanol plants in 2012 as technologies become commercially viable and cellulosic ethanol mandates increase. Until then, pilot plants will likely be in operation, as technologies to produce ethanol from cellulosic biomass are evaluated. In addition, smaller-scale "niche" cellulosic ethanol plants may be built to capture favorable economic opportunities.

Production Processes

Second-generation biofuels generally fall into two main categories based on the processes used to convert the biomass to fuel: biochemical or thermochemical. Biochemical conversion involves the fermentation of sugars from the breakdown of the biomass. Thermochemical involves the "gasification and pyrolysis of biomass material into a synthesis gas or liquid oil for subsequent fermentation or catalysis. The main benefit of gasification and pyrolysis over the biochemical route is that thermochemical processes can more easily convert low-carbohydrate or "non-fermentable" biomass materials such as forest and wood residues to alcohol fuels and can more readily accept a wider variety of feedstocks. However, the thermochemical process does have some drawbacks, such as tar production and clean-up gas procedures that require additional capital investment."[224]

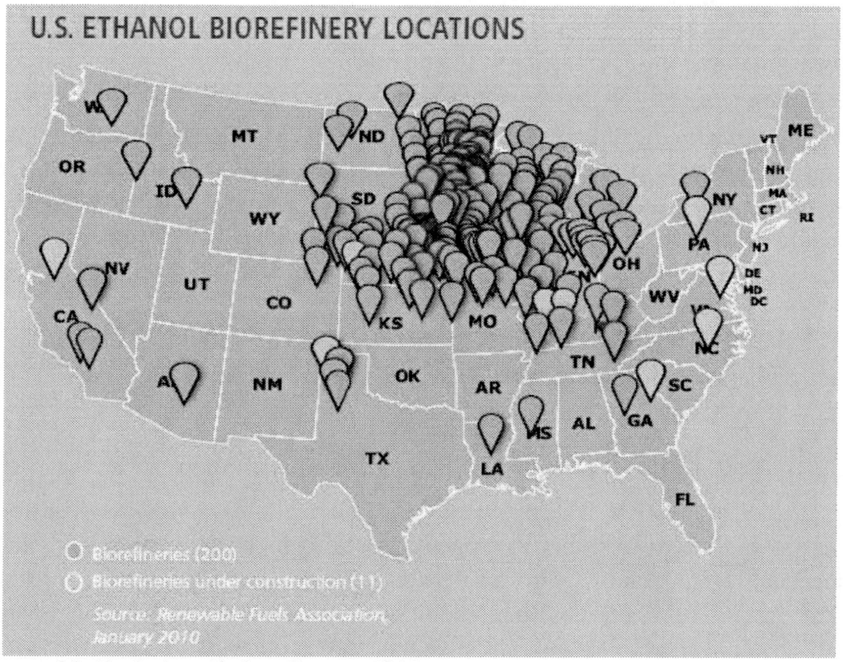

Source: Renewable Fuels Association, January 2010.

Figure B-12. Locations of Current and Planned Ethanol Refineries.

Commercial production of second-generation or cellulosic ethanol is still in its infancy; the predominant and most viable production process has yet to be determined. Planned production plants and R&D are supporting both production pathways.

Biochemical Production

Biochemical cellulosic ethanol production is similar to the production processes for first-generation ethanol. However, since breaking down biomass feedstocks into fermentable sugars is more difficult than for corn, sugar, and other starchbased feedstocks, two key additional steps are required: biomass pretreatment and cellulose hydrolysis. Figure B-13 provides a schematic of the biochemical production process.

The primary steps of the biochemical production process are as follows:

- *Biomass feedstock handling.* The biomass feedstock is reduced in size to improve handling and ultimately, the efficiency of the production process.
- *Biomass pretreatment.* During pretreatment, the hemicellulose in the biomass is converted by hydrolysis into simple sugars using dilute sulfuric acid.
- *Cellulose hydrolysis (sacchrification).* The remaining cellulose in the biomass is hydrolyzed into glucose, aided by cellulase enzymes. This step may be referred to as cellulose saccharification since it produces sugars.
- *Glucose and pentose fermentation.* The glucose and pentose produced by hydrolysis is converted to ethanol through fermentation, resulting from yeast or bacteria feeding on the sugars. The products of the fermentation are ethanol and carbon dioxide.

- *Ethanol recovery.* Ethanol recovery consists of separating ethanol from the fermentation process through distillation and further dehydration to remove remaining water.
- *Lignin utilization.* Byproducts of the biochemical production process, such as lignin, can be burned to produce electricity for reuse in the production process. Typically, lignin generates more electricity than is needed for production, and electricity may be resold.

Thermochemical Production

There are two primary pathways for thermochemical cellulosic ethanol production, gasification, and pyrolysis. These processes are discussed below.

Gasification

Figure B-14 provides a schematic of the gasification pathway for thermochemical cellulosic ethanol production. The process consists of cellulosic biomass feedstock heated with a catalyst and steam (or other hydrogen source) to high temperatures (about 800 °C). The products are H_2, CO, and CO_2 in various concentrations, depending on intended end use. After being purged of impurities and conditioned, this synthetic gas, or "syngas," can subsequently by converted to ethanol via mixed alcohol synthesis using catalytic processes or a fermentation process using microorganisms.

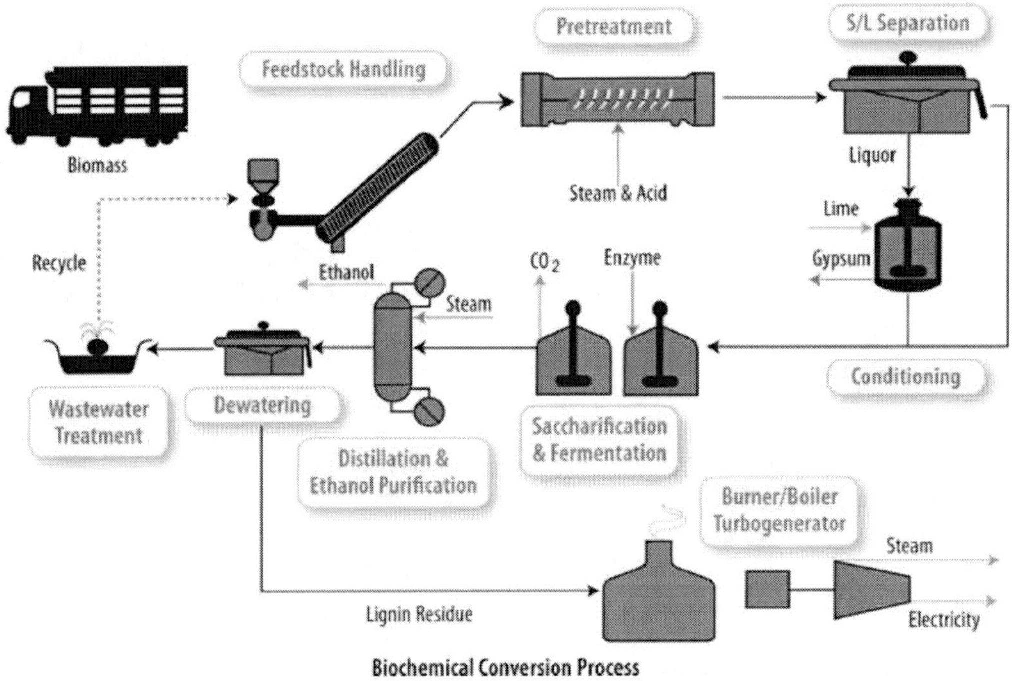

Source: Thomas D. Foust, Andy Aden, Abhijit Dutta, and Steven Phillips, "An economic and environmental comparison of a biochemical and a thermochemical lignocellulosic ethanol conversion processes," *Cellulose* (2009), 16:547–565.

Figure B-13. Biochemical Cellulosic Ethanol Production Process.

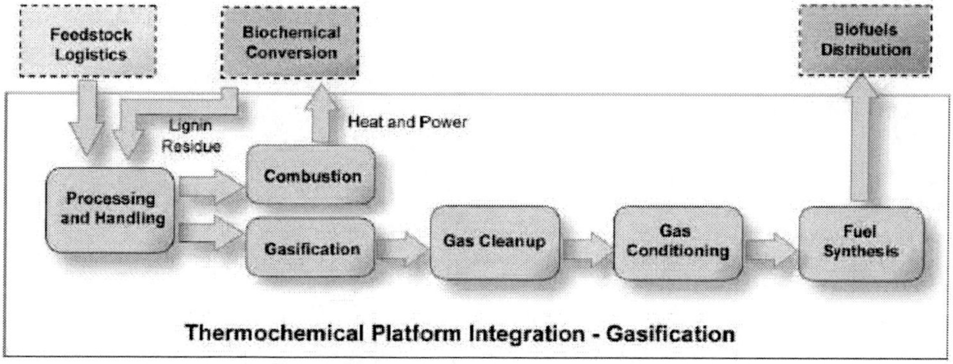

Source: DOE. "Biomass Multi-Year Program Plan." Office of the Biomass Program, Energy Efficiency and Renewable Energy, U.S. DOE, 2010, p. 3–48.

Figure B-14. Thermochemical Gasification Pathway for Biomass to Biofuels.

Pyrolysis

Figure B-15 provides a schematic of the pyrolysis pathway for thermochemical cellulosic ethanol production. The process consists of cellulosic biomass feedstock high temperatures in the absence of oxygen. The carbon-based material decomposes to form bio-oils, biochar, and pyrolysis vapors, which are subsequently cleaned and stabilized for processing into ethanol.

Production Locations

Once feedstock is harvested, collected, stored, and preprocessed, it must be transported to a processing plant for conversion to fuel. There are only limited commercial cellulosic biofuel plants in the United States today. However, cellulosic biofuel production is expanding, with many pilot projects and demonstration plants.

Figure B-16 presents planned and existing locations and capacity for cellulosic biofuel (ethanol and biodiesel) plants. Most production plants are located in the Midwest, the Southern States, and on the Pacific Coast near large cellulosic feedstock sources.

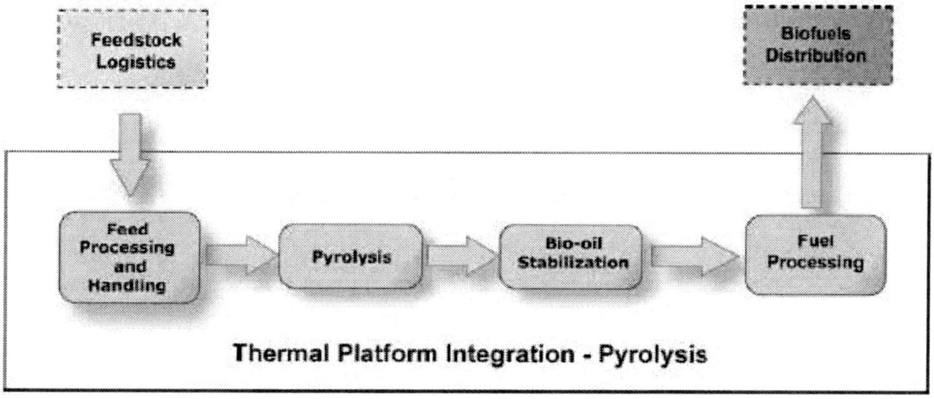

Source: DOE. "Biomass Multi-Year Program Plan." Office of the Biomass Program, Energy Efficiency and Renewable Energy, U.S. DOE, 2010, p. 3–48.

Figure B-15. Thermochemical Pyrolysis Pathway for Biomass to Biofuels.

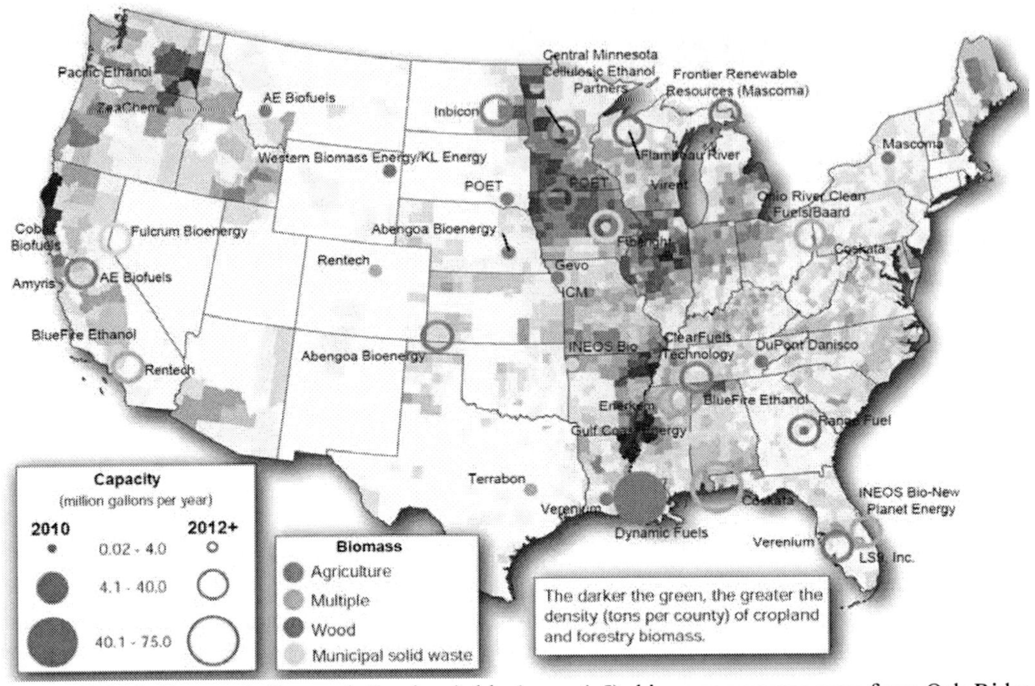

Source: USDA, Economic Research Service (table 1, pp. 4-5); biomass resource map from Oak Ridge National Laboratory (Biomass Research and Development Initiative, December 2008, p. 79)

William T. Coyle, *Next-Generation Biofuels: Near-Term Challenges and Implications for Agriculture*, Economic Research Service, USDA, May 2010.

Figure B-16. Locations of Existing and Planned Cellulosic Biofuel Plants.

Ethanol Production Costs

First-Generation Ethanol

At the June 2010 price of corn ($3.24 per bushel), the net production cost per gallon of ethanol is $1.46.[225] Table B-8 presents a breakdown of the production costs of ethanol.

Ethanol production costs may improve slightly due to recent technological improvements, including fractionation (oil extraction technologies), low heat fermentation, and use of pre-treated or gray-water.[226]

Table B-8. Breakdown of Current Net Production Costs for Ethanol

Production cost component	Production cost ($/gallon)
Corn feedstock	1.16
Natural gas	0.18
Other variable costs	0.21
Fixed costs	0.21
Total production costs	1.77
Sales of byproducts (DDGS)	0.31
Total net production cost	1.46

Second-Generation Ethanol

Current costs for the production of cellulosic ethanol are more than $2 per gallon—Poet, LLC current production cost for cellulosic ethanol produced from corn husks and cobs is $2.35 a gallon.[227] Once plants are operating at economic scale and production processes have been refined, cellulosic ethanol production costs are projected to be less than those of first-generation ethanol. Tables B-9 and B-10 show the breakdown of projected cellulosic ethanol costs for both the biochemical and thermochemical processes. Both are less than the current net production costs of $1.46 per gallon for corn-based ethanol.

Table B-9. Breakdown of Projected Biochemical Cellulosic Ethanol Production Costs (Corn Stover)

Production cost component	Production cost ($/gallon)	Percent
Biomass	0.51	38.5
Feed Handling	Included Above	
Pretreatment/Conditioning	0.25	18.6
Hybrid Hydrolysis & Fermentation	0.10	7.8
Cellulase Enzyme	0.10	7.5
Distillation and Solids Recovery	0.15	11.2
Wastewater Treatment	0.02	1.8
Boiler/Turbogenerator	0.13	9.8
Utilities	0.05	3.8
Storage	0.01	1.0
Total net production cost	1.33	100

Source: Thomas D. Foust, Andy Aden, Abhijit Dutta, and Steven Phillips, "An economic and environmental comparison of a biochemical and a thermochemical lignocellulosic ethanol conversion processes," *Cellulose* (2009), 16:547–565.

Table B-10. Breakdown of Projected Thermochemical Cellulosic Ethanol Production Costs (Wood Chips)

Production cost component	Production cost ($/gallon)	Percent
Feedstock	0.57	46.9
Feed Handling & Drying	0.14	11.8
Gasification	0.11	9.1
Tar Reforming: Acid Gas & Sulfur Removal	0.34	28.0
Alcohol Synthesis Compression	0.10	7.9
Alcohol Synthesis Other	-0.18	-14.6
Alcohol Separation	0.05	3.7
Steam System & Power Generation	0.05	4.4
Cooling Waters & Other Utilities	0.03	2.8
Total net production cost	1.22	100

Source: Thomas D. Foust, Andy Aden, Abhijit Dutta, and Steven Phillips, "An economic and environmental comparison of a biochemical and a thermochemical lignocellulosic ethanol conversion processes," *Cellulose* (2009), 16:547–565.

Cellulosic ethanol production requires higher initial capital investment for biorefinery construction than for first-generation ethanol. As the technology develops and is refined, these investment requirements should decrease. Average capital costs for a 100 million gallon capacity cellulosic biochemical conversion plant are projected to be $320 million, while those of a similarly sized thermochemical conversion plant are around $340 million.[228] These costs are three to four times the capital investment costs for non-cellulosic alternatives.[229]

Ethanol Tax Credit and Implications

In 2004, the American Jobs Creation Act (H.R. 4520, Public Law 108-357), established the VEETC to provide a $0.51 per gallon payment for blending ethanol into the gasoline supply. The Food, Conservation, and Energy Act of 2008 (also referred to as the 2008 Farm Bill) reduced the ethanol tax incentive by 6 cents to $0.45 per gallon.

First-Generation Ethanol Production Tax Credits

With the current tax credit of $0.45 per gallon, first-generation ethanol production costs are reduced to $1.01, and ethanol remains competitive with gasoline as a fuel additive. Note that the tax credit actually is provided directly to ethanol blenders, and it is implied in the comparison of production costs with gasoline. As shown in Figure B-17, the production cost of ethanol rises and falls with the price of corn. The current implied ethanol production price of $1.01 per gallon (with tax credit) is equivalent to gasoline production with crude oil at roughly $39 per barrel. Without the tax credit, the equivalent crude oil price jumps to $57 per barrel.

In addition to the VEETC, the SEPTC (26 *U.S. Code* 40) provides a tax credit for small ethanol production (annual capacity less than 60 million gallons) of $0.10 per gallon on the first 15 million gallons of ethanol produced in a tax year. The VEETC and SEPTC are set to expire on December 31, 2010.

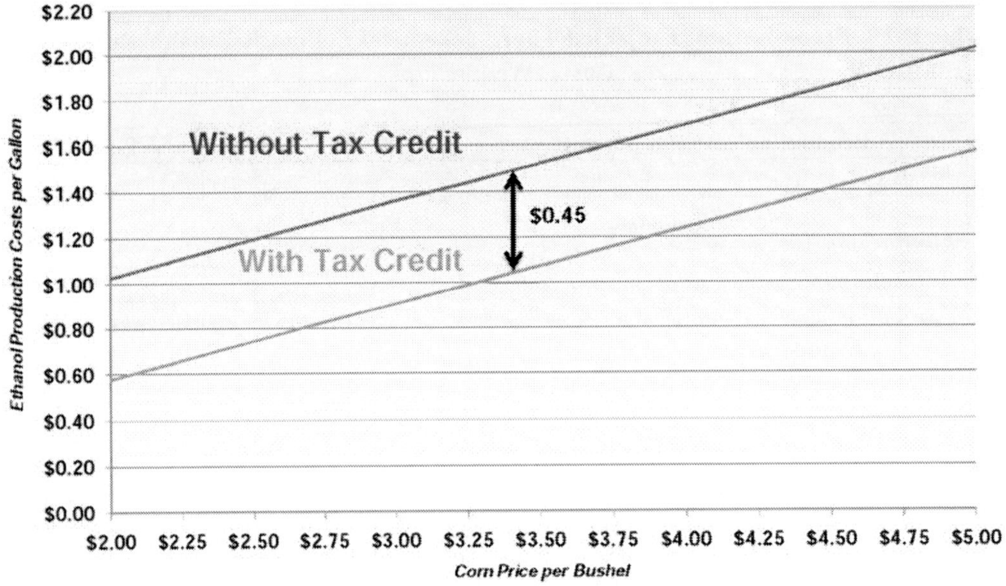

Figure B-17. Net Ethanol Production Costs and Commodity Price of Corn.

Both the VEETC and SEPTC have been major factors in the increase in ethanol production from 3.4 to 10.6 billion gallons from 2004 to 2009. These tax credits provide "a safety net to ethanol producers and blenders against wide fluctuations in oil prices and thereby assures ethanol producers and investors that their product will remain cost competitive over sustained periods of time."[230] Additionally, the tax credits that help support domestic ethanol production is used to fulfill the RFS2 requirements, rather than "enable foreign producers, such as Brazil, to export ethanol directly to the United States competitively."

Second-Generation Ethanol Production Tax Credits

In May 2008, the Food, Conservation, and Energy Act of 2008 (Public Law 110234, Section 15321) established the CBPTC. The CBPTC provides a tax credit of up to $1.01 for producers of cellulosic ethanol. Note that this credit includes the VEETC and the SEPTC credits as components. The CBPTC is set to expire on December 31, 2012. This tax credit is intended to support the early growth of the cellulosic ethanol industry.

Future of Ethanol Tax Credits

On March 25, 2010, Congress introduced legislation, the Grow Renewable Energy from Ethanol Naturally Jobs Act of 2010 (H.R. 4940, GREEN Jobs Act of 2010) extend VEETC, SEPTC, and CBPTC at their current levels through 2015. There is considerable debate on both extending and ending these credits, and the passage of the bill is questionable. If these credits are not extended, there may be some short-term disruptions in the domestic ethanol industry as some higher cost producers close and a potential increase in ethanol imports.

Ethanol Transportation

Denatured ethanol produced at ethanol refineries is transported to bulk terminals before final distribution to retail facilities. Ethanol is mostly transported by rail, barge, and tank truck from the Corn Belt (in the Midwest) to petroleum terminals (which are largely located on the coasts) for mixing and distribution. About 60 percent of corn-based ethanol is transported by rail, 30 percent by truck and 10 percent by barge.[231] It is not commonly shipped in pipelines because it can cause corrosion cracking and can be contaminated when water gets into the line.[232]

There is an incongruity between where ethanol production takes place and feedstock supply with petroleum distribution infrastructure. Currently one-third to one-half of ethanol travels through Chicago, which is a "choke point."[233] As biofuel usage increases, petroleum terminals will need to build accommodations, including upgrades to truck and barge receipt. There are also capacity issues, labor shortages, and public opposition to increased truck traffic.

The challenge to national distribution of ethanol is the limited capacity of the freight rail system, coupled with the lack of dedicated ethanol pipelines.[234] As of July 2010, there was a backlog of 14,930 freight cars, representing slightly less than 1 year of rail car production.[235] We estimate that 27,500 additional freight cars will be required to handle the projected increase in ethanol production by 2020,[236] or roughly 5 percent of all freight cars projected to be produced during this period.

Because no pipeline exists for ethanol transport, moving it to the point of sale is far more expensive per gallon than gasoline. In June 2010, rail tank car transportation costs per gallon for ethanol (from refineries to fueling facilities) ranged from 8 to 26 cents, depending on distance and transportation mode, between two and five times the transportation cost for gasoline, an estimated 3 to 5 cents per gallon.

Pipeline

Ethanol is not currently transported through existing petroleum pipelines, although this mode is the most cost-effective for transporting fuel across significant distances. The primary reasons limiting ethanol in pipelines are (1) ethanol absorbs water and impurities in pipelines, reducing product quality; (2) most ethanol production is not located near pipelines; and (3) individual shipments of ethanol are too small to warrant shipping grade designation.[237] In the foreseeable future, a pipeline dedicated to ethanol transport is unlikely to be constructed, primarily due to high capital requirements relative to potential ethanol pipeline volume. NREL estimates that an ethanol pipeline could cost as much as $1 million per mile.[238]

Tank Barges

Tank barges, at an average cost of $0.72 per ton-mile, are the most cost-effective transportation mode currently available for ethanol.[239] The cost effectiveness stems primarily from scale—one tank barge is the equivalent of 15 rail tank cars or 80 tanker trucks. Barge infrastructure is located near the nation's major waterways, in the Midwest, Northeast, Mid-Atlantic, and Gulf Coast. Although the barge industry has sufficient capacity, ethanol transport via tank barges is limited due to lack of proximity of ethanol refineries to barge terminals and the limited scale of ethanol deliveries.

Rail Tank Cars

At an average cost of 2.24 cents per ton-mile, transportation by rail tank car is three times more costly than via tank barge.[240] Each 30 million gallon capacity ethanol plant requires roughly 150 rail tank cars dedicated to support delivery of its ethanol production, and an additional 150 dedicated freight cars to support feedstock delivery if necessary.[241]

To obtain transportation efficiencies, ethanol is increasingly being transported using unit trains, dedicated freight trains with 75 to 95 tank cars of ethanol.[242] Many ethanol refineries are upgrading their rail yard facilities to handle unit trains. For example, Union Pacific Corporation required Golden Grain Energy, LLC, in Iowa to triple the size of its rail yard in 2006 when its ethanol production increased to a point where unit trains were economically feasible.

The freight rail infrastructure will be stressed by increased ethanol production and sales. *The Wall Street Journal* reports, "As ethanol producers ramp up production, they are straining railroads already taxed by burgeoning shipments of coal, containers, and grain. And they worry that the transportation crunch could make it difficult for ethanol, despite its surge of support in Washington, to compete with energy rivals."[243]

Tanker Trucks

Tanker trucks are the most available but also the most costly ethanol transportation method: its average cost of 26.61 cents per ton-mile is almost 12 times that of rail tank cars.[244] Therefore, tank trucks are primarily used for delivery from the terminal to the retail infrastructure (short distance and lower volume requirements).

Ethanol Blending

Because it tends to separate from gasoline, ethanol is typically blended at distribution terminals, just prior to transportation to retail stations.[245] Therefore, there may be significant storage requirements for ethanol prior to blending and transporting to retail stations. Because ethanol is more corrosive than gasoline, storage tanks must meet unique specifications.

Ethanol Retail Sales

The retail demand for ethanol and available production capacity drive its production. Most of the demand for ethanol in the United States is as an additive (1) to replace MTBE, a suspected carcinogen which has been implicated in the contamination of drinking water, and (2) as a relatively low-cost octane enhancer. As an additive, ethanol is blended with gasoline at 10 percent volume, referred to as E10. EPA is currently (FY10) evaluating whether to increase the maximum blend from E10 to E15.

Ethanol blended as E10 (or gasohol) currently accounts for 99.5 percent of all ethanol used in the United States and is expected to remain near this proportion through 2015. Almost all remaining ethanol production will be blended as E85. E85 retail sales are projected to increase dramatically in 2016, driven by the growth of cellulosic ethanol production coupled with limitations on the volume of ethanol that can be blended into gasohol. EPA estimates that in order to support our forecasts, consumers need to have reasonable access to E85 in 60 percent of the nation by 2022.[246]

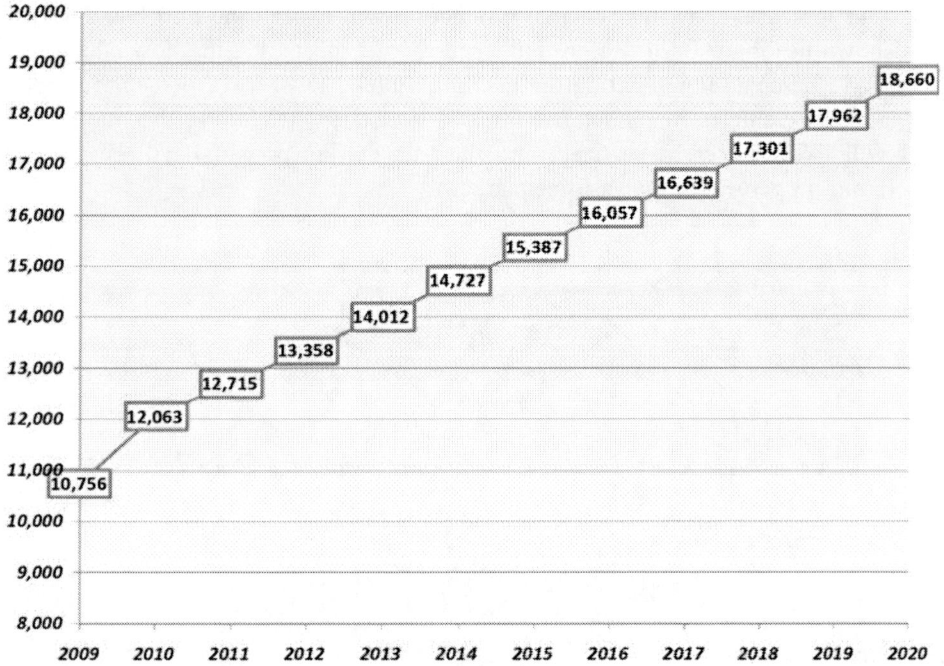

Sources: USDA *Agricultural Projections to 2019*, OCE-2010-1, February 2010; FAPRI at The University of Missouri, *U.S. and World Agricultural Outlook*, 2010; and EPA, *Renewable Fuel Standard Program (RFS2) Regulatory Impact Analysis*, EPA-420-R-10-006, February 2010.

Figure B-18. Current and Projected Retail Sales of Ethanol, 2009–20 (million gallons).

Table B-11. Forecast of Potential Ethanol Retail Sales, 2009–20

	2009	2010	2011	2012	2013	2014	2015	2016	2017	2018	2019	2020
E10, E12, or E15	10,702	12,005	12,654	13,288	13,936	14,646	15,285	14,917	15,153	15,456	15,531	15,431
E85	54.0	57.6	61.3	69.3	75.8	80.5	101	1,139	1,486	1,844	2,431	3,229
Total projected retail sales	10,756	12,063	12,715	13,358	14,012	14,727	15,387	16,056	16,638	17,301	17,962	18,660

Overall Ethanol Retail Sales Forecast

Our forecast for retail ethanol sales was calculated using inputs from three forecast projections: USDA, FAPRI, and EPA. As shown in Figure B-18, we project ethanol retail sales to increase more than 73 percent from 2009 to 2020, from 10,756 to 18,660 million gallons. Table B-11 provides a breakdown of the ethanol retail sales by end-use fuel.

Gasohol (E10 and/or E15) Retail Sales

In 2009, E10 represented 79 percent of all gasoline sold in the United States. As shown in Figure B-19, E10 represented 75 percent more of gasoline sales in 31 states.

As production increases, ethanol retail sales will likely exceed the "E10 blending wall," or the theoretical maximum ethanol that can be blended into the projected demand for gasoline. Our forecasts show that ethanol production will exceed the E10 blending wall in 2013, as shown in Figure B-20. Before this occurs, we expect EPA to allow ethanol blends greater than 15 percent, either setting the maximum blend level at 12 percent or approving E15 but only for vehicles manufactured in 2001 and later. Therefore, the maximum E15 potential will likely never be realized, and the true market potential will be somewhere between 12 and 13 percent of the gasoline market.

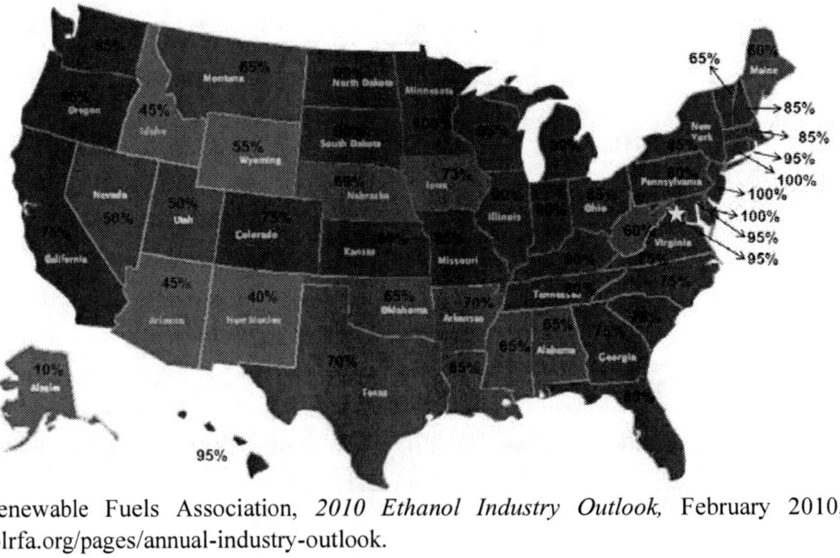

Source: Renewable Fuels Association, *2010 Ethanol Industry Outlook,* February 2010, www. ethanolrfa.org/pages/annual-industry-outlook.

Figure B-19. Estimated Market Penetration of E10, 2009.

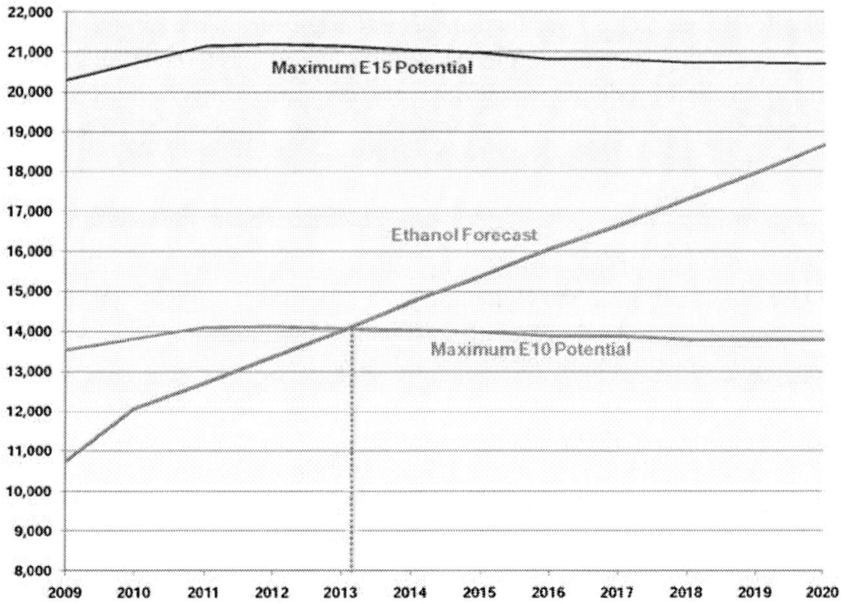

Figure B-20. Market Potential for Ethanol as E10 and E15, 2009–20 (million gallons).

Economics also drive the demand for ethanol blended as E10—the market value of ethanol blended as E10 is far higher than as E85. Since E85 has only 72 percent of the energy content per gallon of gasoline, ethanol blended as E85 must be discounted at the pump to account for its reduced fuel efficiency. However, since ethanol is a relatively cost-effective oxygenate or octane enhancer in gasoline, E10 is sold at the same price as gasoline. Therefore, at the current price of gasoline of $3.00 per gallon, the implied wholesale price of ethanol in E10 is $2.91 per gallon, or 54 percent higher than the implied wholesale price of ethanol in E85 of $1.89 per gallon. Table B-12 shows the calculation of the implied wholesale prices for ethanol in E85 and E10.

Table B-12. Implied Wholesale Prices of E10 and E85 at Current Gasoline Prices

Category	Gasoline	E10	E85
Implied Cost of Ethanol at Retail Pump			
Pump price (per gallon)	$3.00	$3.00	$2.13
Taxes and margin (per gallon)	$(0.60)	$(0.60)	$(0.60)
Implied fuel cost (per gallon)	$2.40	$2.40	$1.53
Neat gasoline actual cost	$2.40		
Sub-octane gasoline actual cost		$2.16	$0.36
Ethanol implied cost		$0.24	$1.17
Implied Ethanol Wholesale Value			
Ethanol revenue per gallon of fuel sold at pump		$0.24	$1.17
Gallons of ethanol		0.10 gallon	0.85 gallon
Value of ethanol		$2.40	$1.38
Volumetric Ethanol Excise Tax Credit (VEETC)		$0.51	$0.51
Ethanol implied wholesale value/price		$2.91	$1.89

Source: Cliff Cook, Marathon Oil, *OPIS Ethanol and Biodiesel Supply Summit: Ethanol Expansion into Growth Markets*, March 2, 2007.

Additionally, the market for gasohol blends is far greater than E85 since gasohol is sold through the same pump and tank systems as gasoline. Therefore, gasohol can be made available at any of the 170,000 retail gasoline stations in the United States. As explained in the next subsection, E85 is limited to slightly more than 1 percent of those retail gasoline stations.

E85 Retail Sales

DOE estimates that E85 is currently available at only 2,096 (or 1.2 percent) of the roughly 170,000 fuel stations in the United States.[247] As shown in Figure B-21, most of these stations are located near production facilities in the Midwest.

The availability of E85 is limited by the following:

- *High infrastructure cost requirements.* Providing E85, as opposed to E10, requires a dedicated dispensing unit, either through retrofitting an existing unit or installing a new one, which typically requires installing an underground storage tank. Costs for installing a new unit may approach $200,000.
- *Dispenser certification issues.* On October 5, 2006, Underwriters Laboratories (UL), an industrial equipment certifying organization, suspended its approval of dispensing equipment for fuels blended with more than 15 percent alcohol over concerns that E85's corrosive nature could result in leaks. After 4 years (in June 2010), UL approved and certified E85 fuel dispensers for two companies, Gilbarco Veeder-Root and Dresser Wayne. We expect this decision to support growth of E85 infrastructure.
- *Low concentration of FFVs.* Approximately 9 million of the roughly 246 million vehicles in the United States are E85 FFVs. Although some areas have higher concentrations of FFVs, generally less than 4 percent of the vehicles passing by gas stations can use E85.

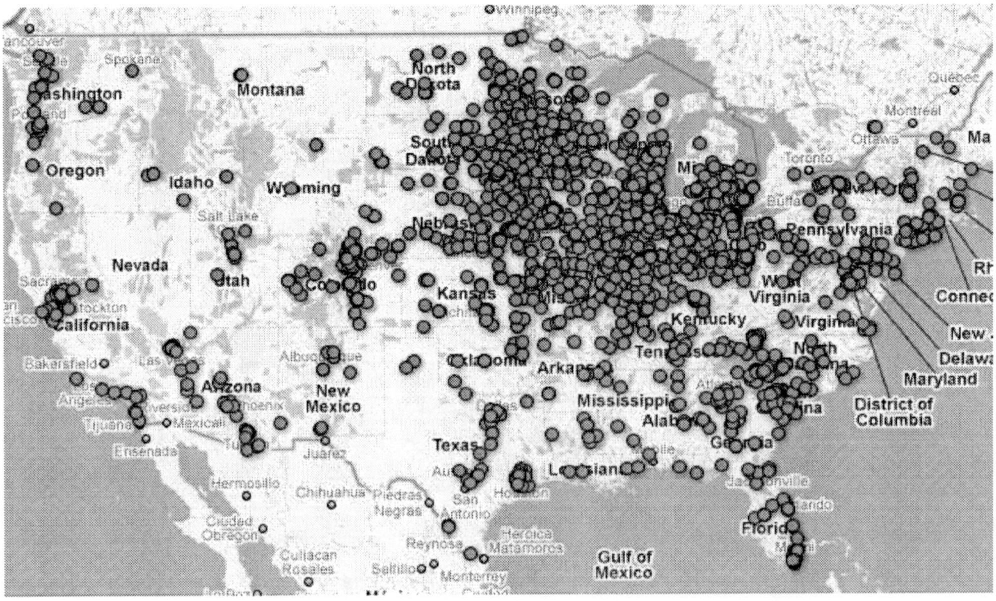

Figure B-21. Location of Commercial E85 Stations, June/July 2010.

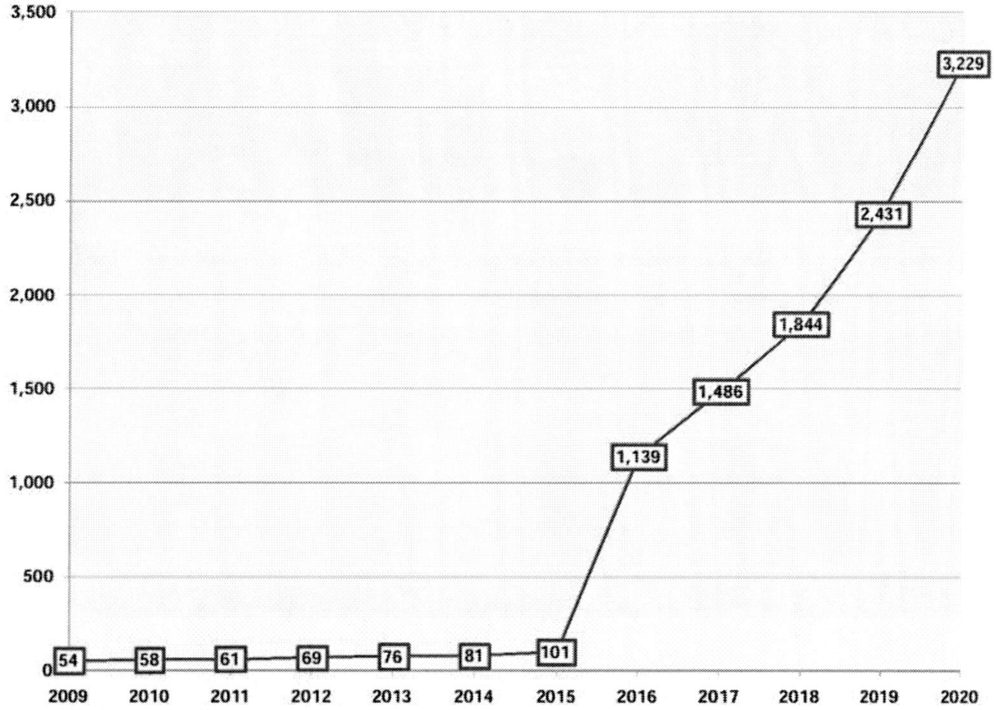

Source: EIA, *Annual Energy Outlook 2010*, DOE/EIA-0383(2010), May 11, 2010.

Figure B-22. Current and Projected Retail Sales of E85, 2009–20 (millions gallons).

- *Price and fuel efficiency of E85 compared with gasoline.* As discussed above, because E85 gets fewer miles per gallon, it typically is sold at a discount to gasoline at the pump. Therefore, blending ethanol as E85 is less profitable than blending as E10.
- *Transportation issues.* As discussed above, most E85 stations are located near ethanol production facilities. Transportation availability, efficiency, and cost limit the national availability of E85 fueling infrastructure.

We forecast retail sales of E85 on the basis of EIA data together with our forecasts of ethanol production and the E10 and E15 markets. As shown in Figure B-22, we project that E85 retail sales will increase almost sixty fold in 2009–20, from 54 to 3,229 million gallons.

Ethanol Prices

Ethanol Commodity Prices

The supply of and demand for ethanol production determine the price of ethanol. When demand and production are in equilibrium, the wholesale price of ethanol is tied to the wholesale price of unleaded gasoline, maintaining a $0.45 per gallon premium based on the tax credit.[248] However, when ethanol demand and supply are unbalanced, ethanol prices begin to deviate from gasoline prices. In cases of excess demand, ethanol prices increase relative to gasoline prices, and vice versa in cases of excess supply.

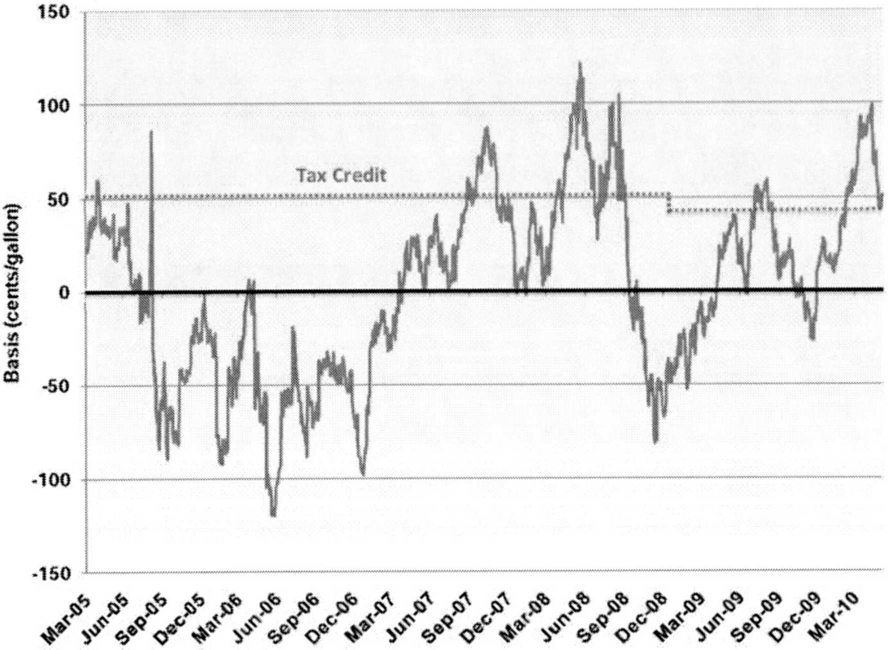

Source: Chicago Board of Trade, *CBOT® Ethanol, Key Charts & Data Updated through June 2007*, June 2007.

Figure B-23. Spread between Ethanol and Unleaded Gasoline Prices.

As shown in Figure B-23, increases in ethanol production capacity in 2006, 2007, and 2009 resulted in a temporary dip in ethanol wholesale prices below unleaded gasoline prices. As demand for ethanol in E10 began to catch up with production capacity, ethanol prices returned to equilibrium in 2010.

Retail E85 Prices

Nationally, retail prices for E85 are lower than for regular unleaded gasoline, primarily to promote E85 use and account for E85's lower energy content (and fuel efficiency). In March 2007, DOE's Clean Cities Program reported that the national average price of E85 was $0.20 (or 9 percent) lower than regular gasoline.[249] With the exception of the Central Atlantic region, E85 was cheaper than gasoline, and the largest price differential ($0.29) was on the West Coast.

Based on its energy content, E85 should be priced at 72 percent of the regular gasoline price—E85 is currently priced at a premium to the consumer. Whether consumers will require a lower E85 price compared with gasoline in the future is unclear.

Appendix C. Biodiesel and Renewable and Cellulosic Diesel Detailed Market Projections

In this appendix, we detail our assessment of the current and future commercial availability of biodiesels. Almost all of the biodiesel produced today represents FAME processes, where vegetable oils or animal fats are used to produce biodiesel through transesterification. The current market for biodiesel has stagnated, primarily due to high retail prices compared with diesel fuel. However, we project the market for biodiesel to increase dramatically over the next 10 years, driven by the RFS2 mandate and the emergence of next-generation renewable and cellulosic diesel fuels.

To forecast the commercial availability of biodiesel, we consider each supply chain stage. The supply chain includes feedstock production and distribution, fuel production, transportation of fuel to bulk facilities for blending with conventional fuels, and finally, transportation to a retail facility for final sale to consumers. Market conditions, costs, and bottlenecks at each of these stages of the supply chain determine the availability of biofuels.

Summary

As shown in Table C-1, we forecast that biodiesel use will increase more than twelvefold to 5,848 million gallons by 2020. Most of this growth will be driven by renewable and cellulosic diesel markets. Throughout the next decade, the primary factor limiting FAME biodiesel availability is retail demand. FAME biodiesel use will continue to be limited by high production costs and retail prices, its chemical properties in colder environments, inconsistent quality, and lack of consumer acceptance.

Over the next 10 years, we forecast emergence of a third-generation biofuel, FAME biodiesel produced from algal feedstocks. High production costs will likely limit this biofuel to less than 100 million gallons through 2020.

Over the next few years, we are forecasting renewable and cellulosic diesel (typically produced from rendered animal fats and reclaimed cooking oils and greases) production to ramp up. Biodiesel produced from these processes meets the fuel specification requirements of ASTM D975 (petroleum diesel fuel), and these drop-in fuels are anticipated to be seamlessly blended into the diesel fuel stream. Since production costs and anticipated retail prices are competitive with conventional diesel, the primary factor limiting renewable and cellulosic diesel availability is production capacity.

Supply Chain Overview

The supply chain for biodiesel is similar to that of ethanol (Figure C-1), except with different feedstocks. First-generation biodiesel includes current FAME biodiesel ("a fuel comprised of mono-alkyl esters of long chain fatty acids derived from vegetable oils or animal fats meeting the requirements of ASTM D 6751 fuel") and renewable diesel (diesel fuel derived from biomass using a thermal depolymerization process that the requirements of ASTM D975 for petroleum diesel fuel). Renewable diesel is included in this category because of its anticipated early use of first-generation oil feedstocks (i.e., animal fats).

FAME biodiesel is produced from soybean oil, other vegetable oils, or animal fats through a process called transesterification, which separates glycerin (or glycerol) from oil. The process generates two primary products, glycerin and biodiesel (alkyl esters[250]). In the next few years, renewable diesel will increasingly emerge in the marketplace. Renewable diesel uses hydrotreating (or thermal depolymerization) processes to produce diesel fuel primarily from first-generation feedstocks, such as fats, waste oils, and greases, that compete for the same feedstock pool as current FAME biodiesel.

Second-generation biodiesel, commonly known as cellulosic diesel, is expected to emerge as renewable fuel within the next few years due to both mandates and improved commercial viability. Cellulosic biodiesel is produced from cellulosic biomass using thermochemical processes, including the Fischer-Tropsch BTL.[251]

Third-generation biodiesel includes FAME biodiesel and diesel fuel produced from advanced feedstocks, such as algal oil. Algal biodiesel offers many potential benefits, including minimal land requirements due to high oil yields, producibility using marginal land, and low quality water inputs (i.e., brine water). However, large quantities of water resources are needed to grow the algal feedstock. These fuels are in their infancy and may not be commercially viable before 2020.

In 2008, approximately 720 million gallons of biodiesel were produced in the United States, almost all from vegetable oils.[252] In 2009, after the worldwide economic recession, production was at 475 million gallons.[253] Figure C-1 shows the rapid growth of production between 2001 and 2008 and the decline in 2009.

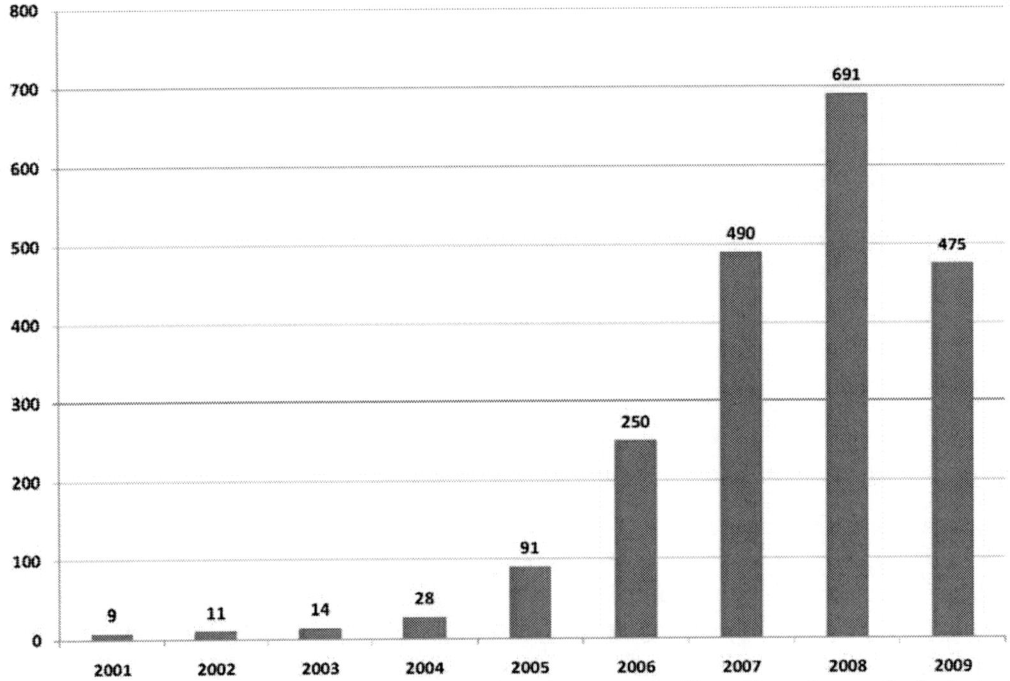

Sources: EIA, *Biodiesel Supply and Consumption, Supplement to the Short-Term Energy Outlook*, April 2009, www.eia.doe.gov/emeu/steo/pub/special/2009_sp_01.pdf on May 6, 2009; FAPRI, *U.S. and World Agricultural Outlook*, January 2010, FAPRI Staff Report 10-FSR 1, ISSN 1534-4533, Iowa State University, University of Missouri-Columbia, Ames, IA (FAPRI6US).

Note: Fuel quantities represent "neat" fuel rather than blended fuel.

Figure C-1. Biodiesel Production 2001–09 (million gallons).

Table C-1. Biodiesel and Renewable and Cellulosic Diesel Forecasts for Supply Chain Stages, 2009–20 (million gallons)

Stage	Type	2009[a]	2010	2011	2012	2013	2014	2015	2016	2017	2018	2019	2020
Feedstock availability	*FAME and Renewable Diesel*												
	Vegetable oils[b]	750	853	956	1,076	1,123	1,126	1,177	1,271	1,315	1,346	1,372	1,382
	Fats and greases[c]	1,205	1,205	1,205	1,205	1,205	1,205	1,205	1,205	1,205	1,205	1,205	1,205
	Total	1,955	2,058	2,161	2,281	2,328	2,331	2,382	2,476	2,520	2,551	2,577	2,587
	Second-Generation Biodiesel (Cellulosic Diesel)												
	Cellulose[d]	33.0 to 99.4 Billion Gallons											
Production capacity	*FAME Biodiesel*												
	Vegetable oils[b]	2,645	3,049	>3,049	>3,049	>3,049	>3,049	>3,049	>3,049	>3,049	>3,049	>3,049	>3,049
	Fats and greases[e,f]	45	51	70	88	107	131	144	167	189	211	233	256
	Algae[f]	0	0	0	0	0	0	0	10	20	40	60	80
	Total	2,690	3,100	>3,119	>3,137	>3,156	>3,180	>3,193	>3,226	>3,258	>3,300	>3,342	>3,385
	Renewable Biodiesel												
	Fats and greases[f]	0	0	44	89	89	126	167	167	167	167	167	167
	Cellulosic Biodiesel												
	Cellulose[f]	0	0	0	111	456	789	1,356	1,922	2,489	3,167	3,844	4,756
	Total	2,690	3,100	>3,163	>3,337	>3,700	>4,095	>4,716	>5,305	>5,893	>6,593	>7,293	>8,227
Projected production and retail sales	*FAME Biodiesel*												
	Vegetable oils[b]	434	598	724	846	914	914	956	1,037	1,081	1,107	1,120	1,108

Table C-1. (Continued)

Stage	Type	2009[a]	2010	2011	2012	2013	2014	2015	2016	2017	2018	2019	2020
	Fats and greases[b,f]	41	46	63	79	96	118	130	150	170	190	210	230
	Algae[f]	0	0	0	0	0	0	0	10	20	40	60	80
	Total	475	644	787	925	1,010	1,032	1,086	1,197	1,271	1,337	1,390	1,418
	Renewable Biodiesel												
	Fats and greases[f]	0	0	40	80	80	113	150	150	150	150	150	150
	Cellulosic Diesel												
	Cellulose[f]	0	0	0	100	410	710	1,220	1,730	2,240	2,850	3,460	4,280
	Total	475	644	827	1,105	1,500	1850	2,456	3,077	3,661	4,337	5,000	5,848

[a] 2009 values represent actual data.

[b] FAPRI, University of Missouri, *U.S. and World Agricultural Outlook,* 2010.

[c] National Biodiesel Board (NBB), *Biodiesel: Feedstock Supply, Achieving 1 Billion Gallons of Biodiesel While Protecting Valuable Feedstocks*, 2010.

[d] EIA, *Annual Energy Outlook 2010*, DOE/EIA-0383(2010), May 11, 2010.

[e] Biomass Research and Development Board, *The Economics of Biomass Feedstocks in the United States: A Review of the Literature*, Occasional Paper No. 1, October 2008.

[f] EPA, *Renewable Fuel Standard Program (RFS2) Regulatory Impact Analysis*, EPA-420-R-10-006, February 2010.

As of June 2009, 173 biodiesel production companies were operating in the United States with annual capacity of 2.69 billion gallons.[254] Approximately 427.8 million gallons of capacity is currently under construction, including the expansion of an existing plant.[255] The 29 relevant companies project completion of new construction by the end of 2010.[256] The estimated annual capacity to produce biodiesel will surpass 3.1 billion gallons by the end of 2010.

Like ethanol, FAME biodiesel cannot be transported through existing multiproduct pipelines. Because of this, according to EIA, "...railroad cars and tanker trucks made from biofuel-compatible materials are needed to transport large volumes of biofuels to market."[257] FAME biodiesel is typically blended as B2 (2 percent biodiesel and 98 percent diesel), B5 (5 percent biodiesel), B20 (20 percent biodiesel), or B100 (pure biodiesel)—most biodiesel is sold as B20.

Renewable and cellulosic diesels meet the diesel fuel specification requirements, and, therefore, can be transported through the same supply chain as conventional diesel. Therefore, these fuels do not raise the same transportation and other supply chain issues as FAME biodiesel. Renewable and cellulosic diesel can be transported through the same pipelines, use the same refueling infrastructure, and be used in the same vehicles as diesel fuel.

Retail sales of FAME biodiesel are projected to increase roughly threefold by 2020. However, biodiesel only represents a very small fraction (less than 1 percent in 2007) of U.S. diesel transportation demand.[258] USDA and FAPRI predict FAME biodiesel production in the United States will increase to more than 1 billion gallons by 2012 or 2013.[259,260] Although not a significant portion of diesel demand, it is nonetheless a growing component of the biofuels picture. The future growth of FAME biodiesel depends on its acceptance by consumers and market demand for diesel. If biodiesel becomes more highly accepted by diesel fleet operators, production may increase further.

We forecast that biodiesel retail sales growth will be driven by the emergence of renewable and cellulosic diesel. By 2015, renewable and cellulosic diesel production volumes will exceed FAME biodiesel. By 2020, these fuels will account for almost 76 percent of biodiesel retail sales.

Biodiesel Feedstocks

Summary

Our forecast for feedstocks for biodiesel production through 2020 is as follows:

- *Vegetable oils will continue as the feedstock for roughly 85 percent of first-generation biodiesel produced.* Soybeans and other vegetable oils will remain the preferred feedstocks for first-generation biodiesel production due to its availability, commodity cost, and economics for biodiesel production.
- *Biodiesel produced from fats, waste oils, and greases will grow almost tenfold by 2020.* Production costs for biodiesel using fats, waste oils, and greases are often lower than for vegetable oil feedstocks. "Much of this biodiesel production, however, seems to rely on niches of feedstock availability and market outlets."[261] EPA projects that by 2020, 230 million gallons per year of rendered or reclaimed fats will be

processed by firstgeneration biodiesel plans and 150 million gallons per year will be used by renewable diesel facilities.[262]

- *The cellulosic pathway will be commercially viable after 2012 and grow rapidly through 2020.* Driven by the RFS2, we project diesel produced from cellulosic feedstocks to begin in 2012 and grow to over 5 billion gallons by 2020. Dedicated energy crops (switchgrass, energy cane, and biomass sourghum) are forecast to comprise roughly 65 percent of the cellulosic feedstocks, with 21 percent crop residues (corn stover and straw) and 14 percent woody biomass (primarily logging residues).
- *Algal FAME biodiesel will emerge in 2016, but will be limited in commercial viability until after 2022.* EPA projects limited (100 million gallons) commercially viable FAME biodiesel production from algal feedstocks by 2022 (EPA does not forecast any production of ethanol or diesel fuel from algal feedstocks before 2022).[263]

First-Generation Biodiesel

Almost all of the biodiesel produced today is considered a first-generation biofuel, or a biofuel generated from crops or animal fats that can also be used as food and feed. The primary potential first-generation feedstocks for biodiesel production in the United States through 2020 are as follows:

- *Soybean oils.* Until 2006, soybean oil represented the feedstock for almost all biodiesel production in the United States—97 percent of biodiesel production in 2006 was from soybean oil. Between FY07 and FY09, soybean oil dropped to 45 percent of biodiesel production, primarily due to a spike in soybean prices during this period. We forecast soybean oil feedstock to comprise between 48 and 53 percent of first-generation biodiesel production through 2020.
- *Other vegetable oils.* Corn oil (including oil extracted during ethanol production) and canola oil are the primary other vegetable oils used for biodiesel production. In 2009, other vegetable oils represented roughly 39 percent of biodiesel production. Other vegetable oil feedstocks are forecast to decrease from 46 percent of first-generation biodiesel production in 2010 to 30 percent in 2020.
- *Animal fats and restaurant grease.* Rendered animal fats and reclaimed cooking oils and greases represent the other major source of biodiesel feedstock. In 2009, animal fats and restaurant grease represented roughly 9 percent of biodiesel production. We forecast these feedstocks to increase to comprise more than 17 percent of first-generation biodiesel production through 2020.

Feedstock Yields and Costs

The economics of biodiesel production primarily depend on the cost of the feedstock. Feedstocks costs are the largest component of biodiesel production costs—the Iowa State Center for Industrial Research and Service estimates that feedstock costs constitute 72 percent of total biodiesel production costs.[264] Table C-2 and Figure C-2 compare wholesale production costs for first-generation biodiesel produced from different feedstocks.

Table C-2. Comparison of Biodiesel Production Costs by Feedstock

Feedstock	Wholesale feedstock price ($ per pound)	Wholesale feedstock price ($ per gallon of biodiesel)
Soybean oil	0.3683	2.76
Corn oil	0.3696	2.77
Canola oil	0.4461	3.35
Cottonseed oil	0.4212	3.16
Peanut oil	0.6207	4.66
Sunflower oil	0.5565	4.17
Tallow	0.2775	2.08
Yellow grease	0.2400	1.83

Sources: Food and Agricultural Policy Research Institute (FAPRI), *U.S. and World Agricultural Outlook*, January 2010, FAPRI Staff Report 10-FSR 1, ISSN 1534-4533, Iowa State University, University of Missouri-Columbia, Ames, Iowa (FAPRI6US); Fred Wellons, National Renderers Association, *Transforming Animal Fats and Used Cooking Oils into Green Fuels–Technology that Works*, March 16, 2010.

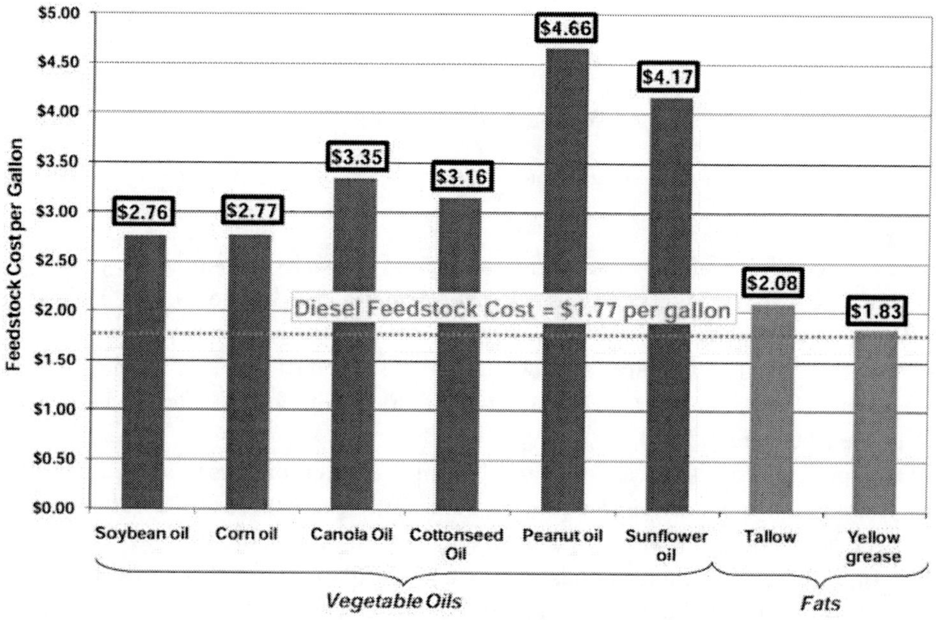

Figure C-2. Comparison of Biodiesel Production Costs by Feedstock.

Of the vegetable oil feedstocks (soybean, corn, canola, cottonseed, peanut, and sunflower oils), soybean and corn are the most cost competitive for biodiesel production. Although production costs from fats are lower than for vegetable oils, supply availability and potential fuel quality issues limit the potential from these feedstocks.

With the exception of yellow grease, feedstock costs per gallon of biodiesel are significantly higher than feedstock cost per gallon of diesel produced. Currently, only high subsidy levels enable biodiesel to remain cost competitive with diesel in retail markets.

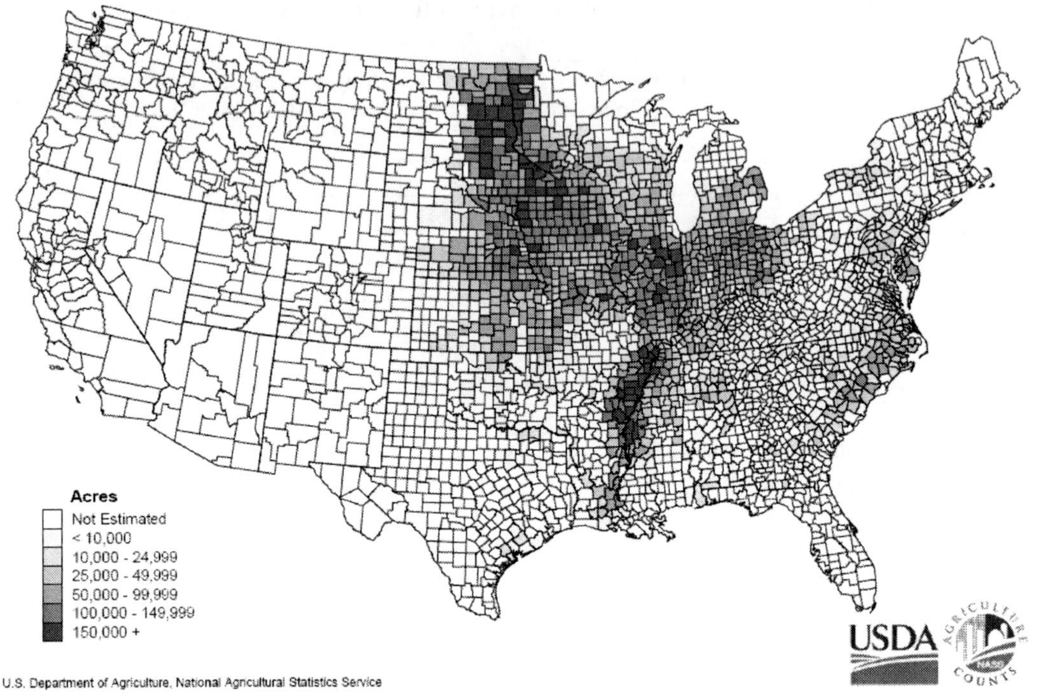

Source: USDA, National Agricultural Statistics Service, August 2010.

Figure C-3. Planted Acres of Soybeans by County (2009).

Soybean Oil

USDA reported that in 2009, 75.7 million acres of soybeans were planted, yielding 2,967 million bushels.[265] As shown in Figure C-3, the majority of this acreage is located in the Midwest, the Mid-Atlantic states, and along the Mississippi.

Annual crop yields have increased over the last 10 years from 38.1 bushels per acre in 2000 to 44 in 2010.[266] USDA projects that advances in technology will further increase yields to 46.5 bushels per acre in 2020.[267]

FAPRI projects that soybean oil available annually for biodiesel production will rise sharply from 1,904 to 5,412 million pounds from 2009 to 2020. This rise reflects projected soybean oil production above a "baseline" demand for nonbiodiesel uses of soybean oil between 14.2 and 15.7 billion pounds from 2009 to 2020. From this growth in available soybean oil, we project that the potential biodiesel that can be produced from soybean oil will increase almost threefold, from 254 to 722 thousand gallons between 2009 and 2020 (Table C-3 and Figure C-4).

The EIA has explained why soybean oil dominates as a feedstock for biodiesel production:

> Soy is a versatile, nitrogen-fixing crop that yields oil and food for humans and livestock. Soybean meal is of higher market value than soy oil. Consequently, soy oil is a low-priced byproduct available in relatively large volumes. Currently, it is a cheaper virgin feedstock than other oilseeds. The processing and distribution infrastructure for

soybeans is already in place, with more capacity being added as more biodiesel production facilities come online.[268]

Table C-3. Forecast of Potential Biodiesel Production from Soybeans, 2009–20

	2009	2010	2011	2012	2013	2014	2015	2016	2017	2018	2019	2020
Million acres planted	75.7	77.5	76.5	73.5	74.5	75.5	76.0	76.0	76.0	76.0	76.0	76.0
Yield (bushels per acre)	39.7	43.3	42.8	43.2	43.6	44.4	44.4	44.9	45.3	45.7	46.4	46.5
Soybeans produced[a]	2,967	3,319	3,230	3,130	3,205	3,280	3,330	3,370	3,400	3,430	3,460	3,490
Soybean oil production[b]	18,746	19,072	20,301	20,714	21,150	21,490	21,827	22,212	22,612	22,988	23,358	23,742
Soybean oil for food use[b]	14,481	14,222	14,562	14,459	14,608	14,792	14,934	15,041	15,171	15,309	15,474	15,666
Soybean oil for biodiesel use[b]	1,904	2,335	3,077	3,655	3,911	3,817	4,080	4,650	4,982	5,210	5,380	5,412
Potential biodiesel production[c]	254	311	410	487	522	509	544	620	664	695	717	722

[a] Millions of bushels.
[b] Millions of pounds.
[c] Thousands of gallons.

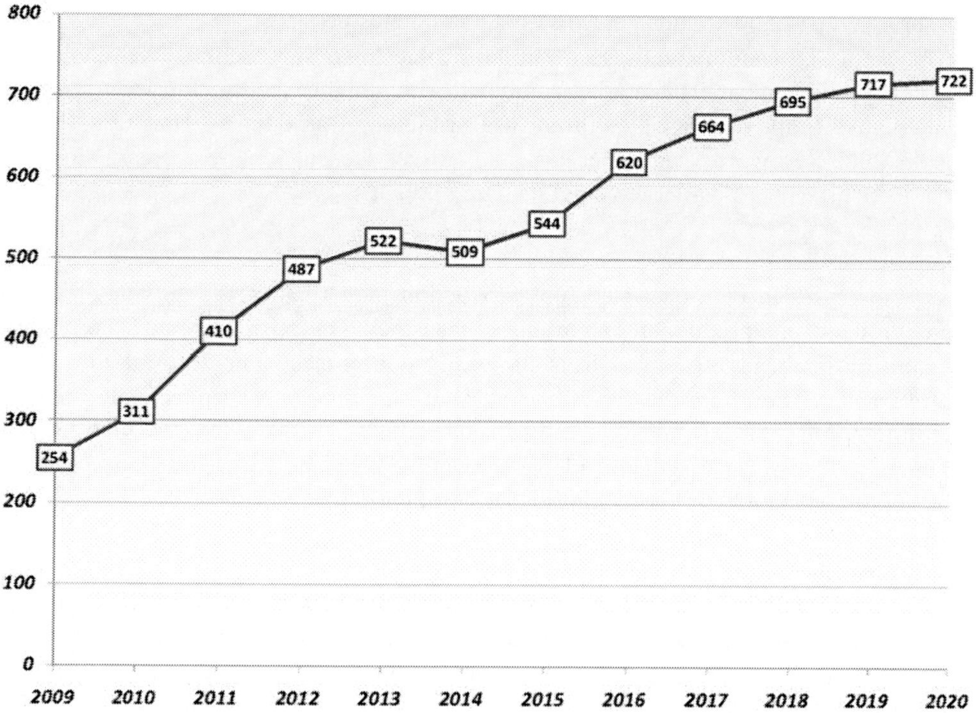

Figure C-4. Forecast of Potential Biodiesel Production from Soybeans, 2009–20 (thousand gallons).

Production of biodiesel from soybeans is not expected to significantly impact the food markets in the near term. However, if the primary feedstock for biodiesel production once again becomes soybean oil, and demand for biodiesel continues to grow, high demand for biodiesel soybean oil may raise the wholesale price of soybeans toward the end of the decade.

Increased biodiesel production has already affected the market for glycerin. Roughly 10 pounds of crude glycerin is generated as a co-product for every 100 pounds of biodiesel production. The amount of glycerin generated by the biodiesel industry has resulted in substantial oversupply (amounting to most of the glycerin produced in the United States) and a drop in the price of crude glycerin.[269]

Other Vegetable Oils

The primary vegetable oils other than soybean oil used for biodiesel production include canola oil and "oil extracted from corn or its fermentation co-products in the dry mill ethanol production process." Table C-4 shows our forecast of the potential biodiesel production from other vegetable oils.

Canola Oil

Canola is a type of rapeseed that contains roughly 40 percent oil. USDA reports that in 2009, 1.01 million acres of canola seed were planted, yielding 1,058 million pounds of canola oil. U.S. demand for canola exceeds domestic production; therefore, more than 61 percent of canola oil used in the United States is imported, primarily from Canada. Canola is grown during the spring and winter, with yields for the winter crop significantly higher than those for the spring crop (1,500 pounds per acre vs. 3,500 pounds per acre).[270]

Most canola is manufactured as an edible oil due to its low saturated fat and high in omega-3 fatty acid content. This market combined with a higher feedstock price compared to other alternative limits canola oil's current use in biodiesel production (roughly 44 million gallons in 2009). However, future use as a biodiesel feedstock may increase due to its potential to be grown as a winter crop and its high oil yields.

Corn Oil

Large increases in corn-based ethanol production may support the growth of corn oil generated from ethanol production as a biodiesel feedstock. For existing wet mill corn ethanol plants, corn oil is already generated from the production process. For dry mill corn ethanol plants, "extraction of oil from the thin stillage or DGS streams is a proven technology that can be retrofitted into existing plants relatively cheaply."[271] EPA estimates that by 2022 approximately 70 percent of ethanol plants will implement corn oil extraction systems by 2022, generating up to 680 million gallons per year of corn oil.[272]

Fats and Greases

Rendered animal fats and reclaimed cooking oils and greases provide a large potential feedstock for biodiesel production through 2020. The NBB estimates that the annual volumes of animal fats and restaurant greases available for biodiesel production are 905 million gallons and 300 million gallons, respectively.

Second-Generation (Cellulosic) Diesel

Second-generation biodiesel includes cellulosic diesel produced from cellulose hemicellulose or lignin. Cellulosic feedstocks include corn stover, timber wastes, and dedicated energy crops such as switchgrass. Technology enabling the production of cellulosic diesel from cellulosic materials is still in its early stages, but the market is expected to grow dramatically over the next 10 years, driven primarily through the RFS2.

We forecast the availability of cellulosic feedstocks to support the production of between 33.0 and 99.4 billion gallons of cellulosic biofuels. Discussion of the availability of cellulosic feedstocks is presented in Chapter 6.

Third-Generation (Algal) Biodiesel

Third-generation biodiesel includes any biodiesel or diesel fuel produced from algal feedstocks. Algae used for biofuel production include diatoms, green algae, golden-brown algae, prymnesiophytes, eustigmatophytes, and cyanobacteria, all single-celled algae species that grow rapidly to form biomass containing oils and residuals used for fuel production.[273]

Algae offers an advantage as a biofuel feedstock since oil yield per acre from algae can be significantly higher than those of other potential biofuel feedstocks, as shown in Table C-5.

As shown in Figure C-5, algal oils can be used to produce a variety of biofuels, including FAME biodiesel, diesel, ethanol, and biobutanol, depending on conversion process. EPA forecasts that only the FAME biodiesel production pathway will be commercially viable before 2020.

Algal Oils Productivity

The main external factors affecting algae-based oil production are climate, water and nutrients source, and CO_2 source. Being photosynthetic, most algae strains require sufficient access to sunlight during the producing period (typically yearround) in order to be productive. Other climate factors such as temperature, precipitation, and evaporation also impact the productivity of algae. One of the benefits of algae is that wastewater or saline or brackish water can be used to effectively grow algal cultures since the algae can absorb the nutrients this type of water contains. A final consideration is that algae require a CO_2-rich environment in order to achieve optimal growth.

Table C-4. Forecast of Potential Biodiesel Production from Other Vegetable Oils, 2009–20 (million gallsons)

	2009	2010	2011	2012	2013	2014	2015	2016	2017	2018	2019	2020
Canola oil	44	46	41	49	48	48	50	53	51	49	48	49
Corn oil	352	392	412	428	444	460	469	478	484	491	498	500
Other oils	100	105	93	111	109	109	114	120	116	111	109	111
Total potential other vegetable oils	496	542	546	589	601	617	633	651	651	651	655	660

Table C-5. Biofuel Yields from Various Feedstocks

Crop	Oil yield (gal/acre/year)[a]
Corn	18
Soybean	48
Canola	127
Jatropha	202
Coconut	287
Oil palm	635
Algae	1,000–4,000

Sources: DOE Biomass Program, *National Algal Biofuels Technology Roadmap*, 2010; Chisti, Yusuf, *Biodiesel from Microalgae*, Biotechnology Advances, February 2007.

[a] Converted from liters/hectare.

Climate

An abundant source of sunlight is one of the major contributing factors toward algae growth. Figure C-6 displays the annual average solar radiation throughout the United States. This map indicates that the U.S. Southwest, especially Arizona, Eastern California, and New Mexico, presents the greatest opportunity for the development of the algal biofuels market.

Precipitation and evaporation levels can also impact algae growth. Areas of the United States that not only have an abundant source of sunlight, but also relatively high levels of precipitation can ensure a steady, long-term source of water for the growing algae. Higher levels of evaporation in a region, while having very little impact on closed algae growth systems, can be detrimental to open systems, such as raceway farms. Regional severe weather conditions need also be considered in the siting process in order to ensure limited exposure to potential system damage, either through equipment damage of a closed system or through contamination of an open one.

Water and Nutrients

One of the most advantageous aspects of algae is that algal cultures do not require fresh water to grow well. Algae can grow well in saline or brackish water as well as in wastewater. This aspect of algae opens up multiple economic opportunities for algae farms, allowing for the growth of algal cultures in areas where most other plants or crops could not potentially grow, thus limiting competition for land resources.

Saline aquifers are one potential source for economic water supplies. As long as the depth of these aquifers is not economically prohibitive, extraction should not be a significant issue since the water is used for little else. Additionally, locations along the coast of the United States would offer the same ready access to saline water.

Wastewater presents a great potential source of water for algae farms. Algae can be used to treat wastewater containing organic matter, excess nutrients, metals, synthetic organic compounds, and potentially endocrine disrupting compounds.[274] Thus, a potential opportunity exists in the co-location of algae farms with wastewater facilities in the United States. Figure C-7 shows the location of all wastewater treatment facilities in the United States. Algal farms could use this wastewater as input and not only economically grow algae but also treat the wastewater effectively.

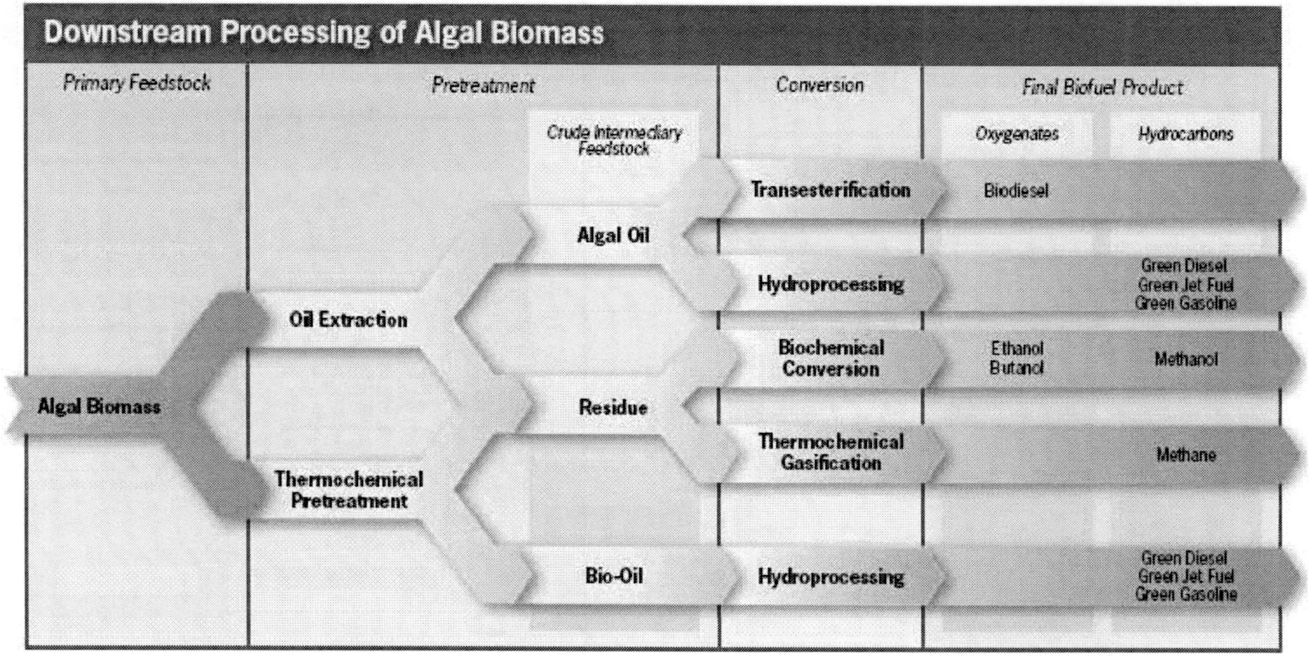

Source: Catie Ryan, Natural Resources Defense Council and Terrapin Bright Green, LLC, *Cultivating Clean Energy: The Promise of Algae Biofuels*, October 2009.

Figure C-5. Pathways for the Downstream Processing of Algal Biomass.

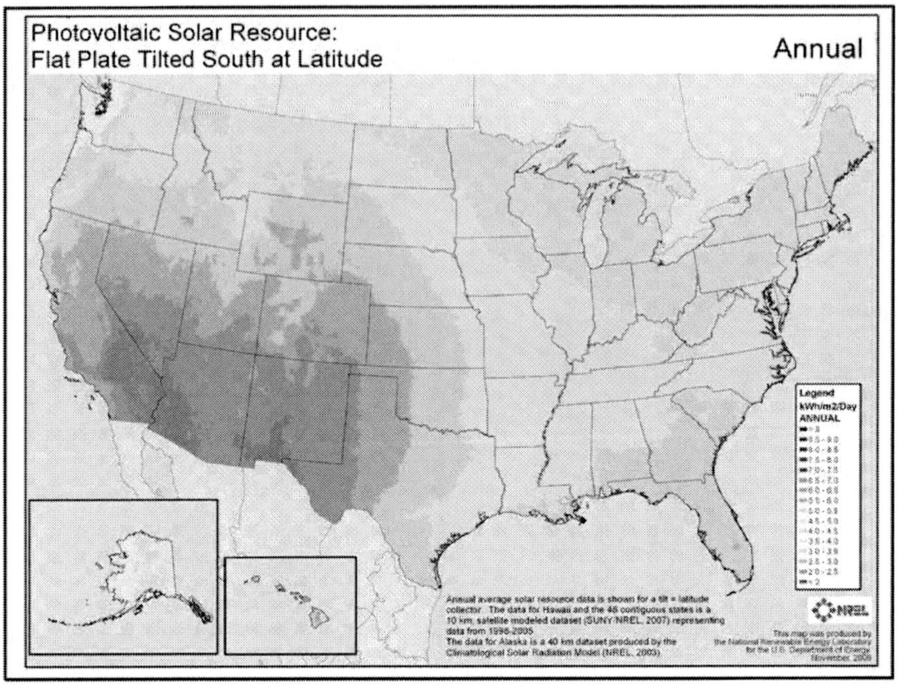

Source: NREL, Solar Maps, November 2008.

Figure C-6. U.S. Average Photovoltaic Solar Resource.

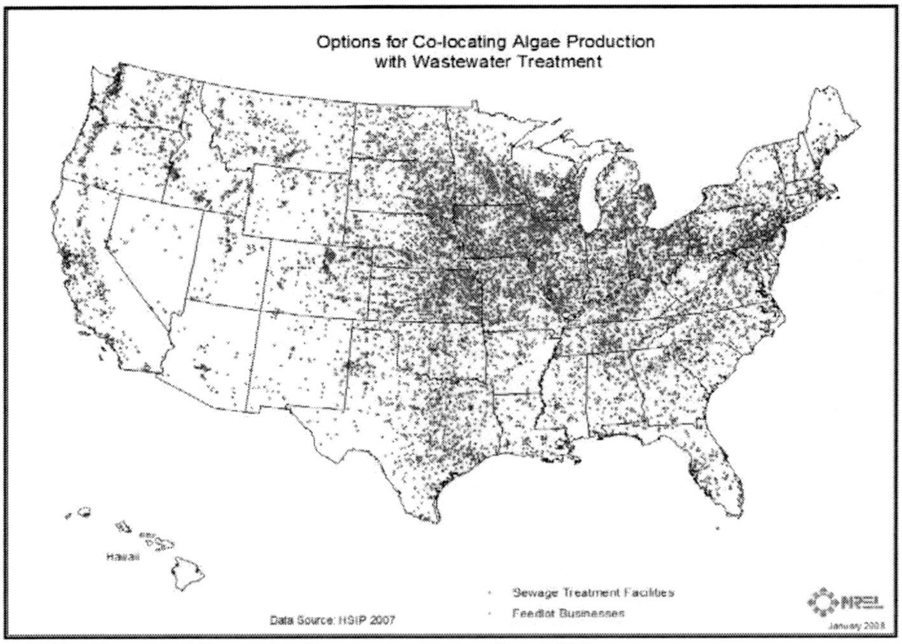

Source: NREL, Homeland Security Infrastructure Protection, 2007.

Figure C-7. Wastewater Treatment Facilities within the United States.

CO_2

In order to ensure optimal productivity on an algae farm, an abundant source of CO_2 is required to enhance growth. This property of algae offers another potential opportunity since major CO_2 emitters could use algae as a means of sequestering their emissions. In 2009, the United States emitted around 5.4 billion metric tons of CO_2, with emissions from the power sector alone accounting for over 2.1 billion metric tons—about 39 percent—of these emissions.[275] If only half of these emissions were captured and used to enhance algae biomass growth almost 200 to 600 million gallons of algal-based biofuels could be produced annually, accounting for between 50 percent and 150 percent of the current amount of diesel fuel consumed in the transportation sector.[276]

Figure C-8 is a map of all major stationary sources of CO_2 in the United States, which represent the greatest potential sources of CO_2 for algae farms. Although there are very few major sources available in Arizona, Southeastern California, and New Mexico—all prime areas for algae growth based upon climate—Southwest Texas has multiple natural gas treatment plants that could be utilized as sources of CO_2 for the market.

Potential Issues

Although the production of algal-based biofuels can be accomplished in the United States, there are some challenges and issues that may make it harder than it appears to be. For one, as mentioned previously, although the United States does have a significant region where optimal algae growth can be achieved, these areas, as demonstrated in Figures C-7 and C-8, do not, for the most part, have easy access to environmentally beneficial sources of water and CO_2.

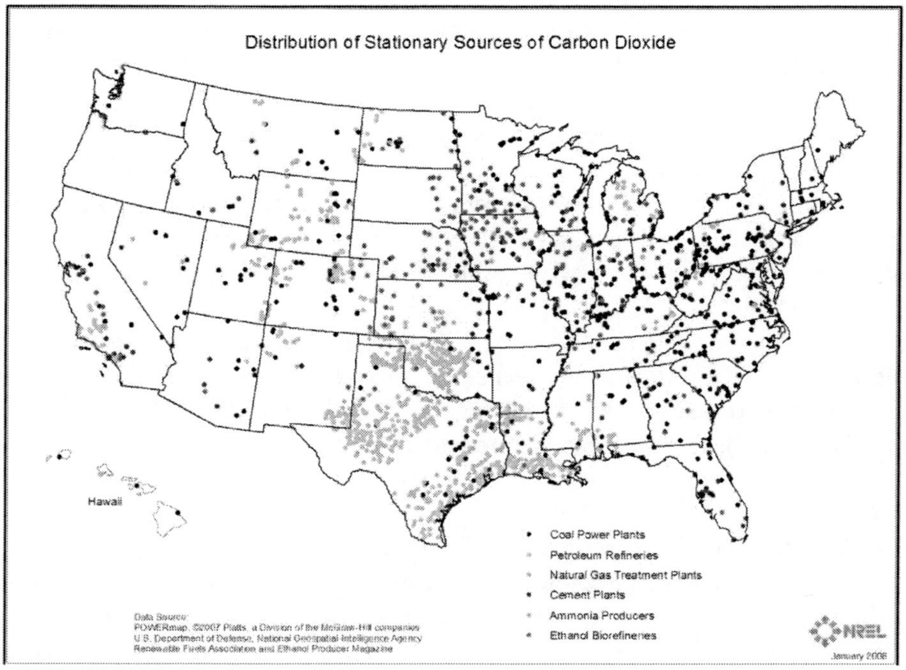

Source: NREL POWERmap, January 2008.

Figure C-8. Major CO_2 Emitters in the United States.

The use of impaired water also presents some difficulties for the production of algae whether open ponds or photobioreactors are used to grow the algae. With open ponds, the salinity of the ponds could increase over time through the use of only saline or brackish water. At some point the ponds would either need to be desalinated or drained and re-filled. Photobioreactors, which are much more capital intensive than open ponds, could most likely run using wastewater as high-value algae products are required to make them economical.[277]

There are a number of potential issues regarding CO_2 sources. Not all CO_2 emissions can be used in algae growth as some of the materials in the emissions are damaging to algae. For example, the sulfur compounds found in coal-fired flue gas are toxic to algae.[278] Also, since algae can only grow during the day, when the sun is shining on it, algae could not be used to sequester emissions from power plants during the evening. And finally, there is a mentality issue within the power generation industry. Power generators see algae simply as a source for CO_2 emissions reductions and do not see algal fuels as a potential revenue stream. Such a mentality could inhibit the growth of co-located algae farms and power plants since the potential revenues from fuel sales are not taken into consideration when doing an economic study concerning the possibility.

Biodiesel Production

First-Generation Biodiesel

Biodiesel annual production capacity currently exceeds the amount of biodiesel produced.[279] The NBB reports that as of June 2009, the annual operational capacity of the biodiesel industry in the United States was 2.69 billion gallons (173 companies). Approximately 427.8 million gallons of new capacity is under construction, and 29 companies project completion of new construction by the end of 2010.[280]

As shown in Figure C-9, currently less than half of all biodiesel refinery capacity is strictly tied to soybean oil as the primary feedstock—approximately 1.1 billion gallons (or 41 percent) of the 2.69 billion gallons of total capacity. It is likely that the majority of capacity, made up of multi-feedstock refineries (52 percent, or 1.4 billion gallons), also currently uses soybean oil as a primary feedstock, although they are able to process vegetable oils, animal fats, and restaurant grease. Less than 200 million gallons of capacity is devoted to canola oil and other oils and fats.[281]

As shown in Figure C-10, most biodiesel refineries are located in the Midwest, Southeast, and Mid-Atlantic, near the primary feedstock sources. There are some exceptions, as biodiesel refineries are also concentrated in areas such as California and Texas. It appears that expanded plants will be located in similar areas as well.

FAME Process

Most first-generation biodiesel is currently produced using the FAME process, in which plant oils and animal fats (triglycerides) are chemically reacted with an alcohol (primarily methanol). The primary reaction is catalyzed using sodium or potassium hydroxide to create an alkaline pH, and occurs at low temperature (150°F) and pressure (20 psi). The production process is very efficient (98 percent conversion) and creates glycerin (used in pharmaceuticals and cosmetics) as a byproduct. Figure C-11 provides a schematic of the FAME biodiesel production process.

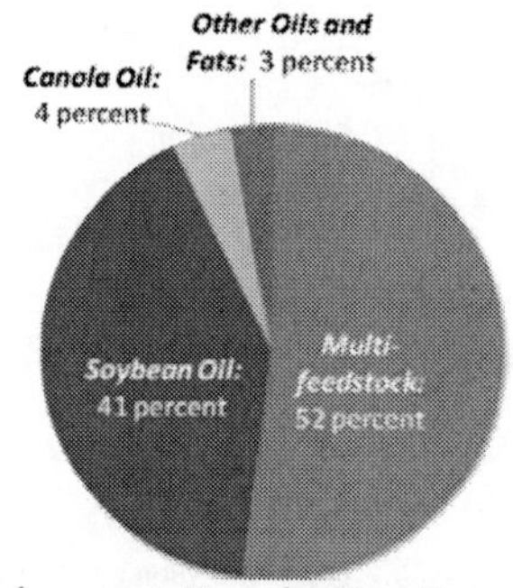

Source: Biomass Research and Development Board, *The Economics of Biomass Feedstocks in the United States: A Review of the Literature*, Occasional Paper No. 1, October 2008.

Figure C-9. Capacity of Current Biodiesel Production Plants by Feedstock.

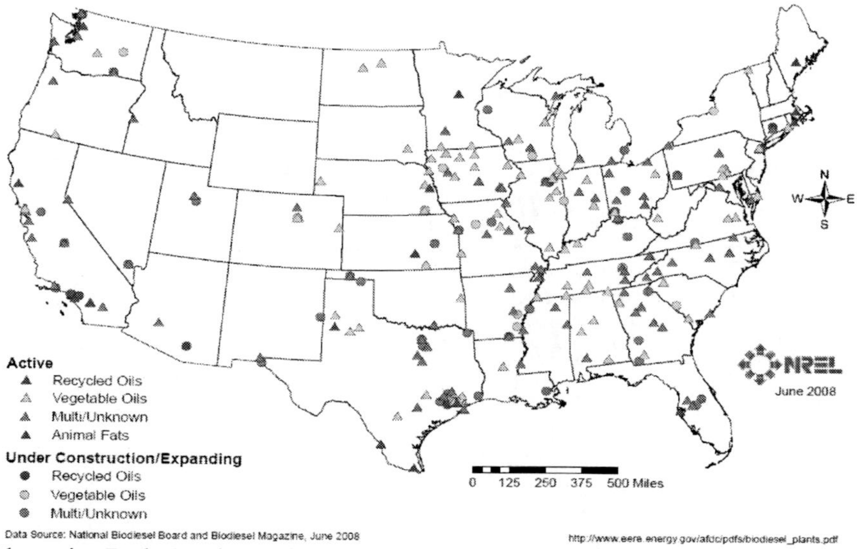

Source: Alternative Fuels & Advanced Vehicles Data Center, www.afdc.energy.gov.

Figure C-10. Locations of Current and Planned Biodiesel Production Plants (June 2008).

Renewable Diesel

Renewable diesel is an emerging biodiesel production process where vegetable oils and animal fats are converted into diesel fuel using thermal depolymerization (a process used in petroleum refining). "The process uses hydrogen and catalyst to remove oxygen from the triglyceride molecules in the feedstocks oils via a decarboxylation and hydro-oxygenation reaction, yielding some light petroleum products and water as co-products. The reactions can also saturate the olefin bonds in the feedstock oils, converting them to paraffins; additional steps can also be taken to isomerize a portion of the paraffins to create fuels with varying properties."[282] Although additional processes can produce a variety of petroleum products, EPA assumes roughly 90 percent of yield to diesel with the remainder divided between light fuel gas and naphtha.[283] A schematic of the renewable diesel production process is provided in Figure C-12.

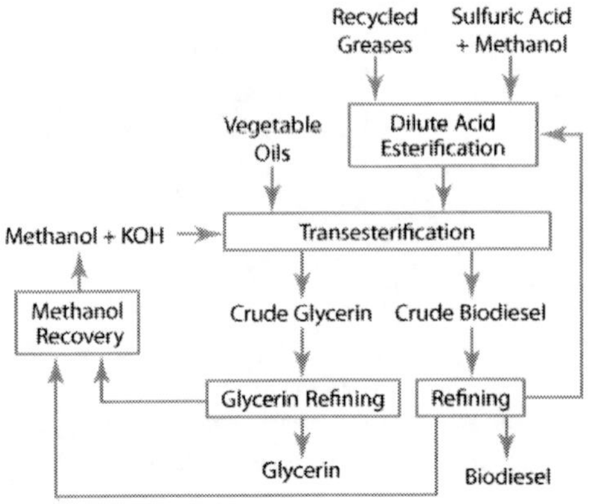

Source: Alternative Fuels & Advanced Vehicles Data Center, *Biodiesel Production*, www.afdc.energy.gov www.afdc.energy.gov/afdc/fuels/biodiesel_production.html.

Figure C-11. FAME Biodiesel Production Process.

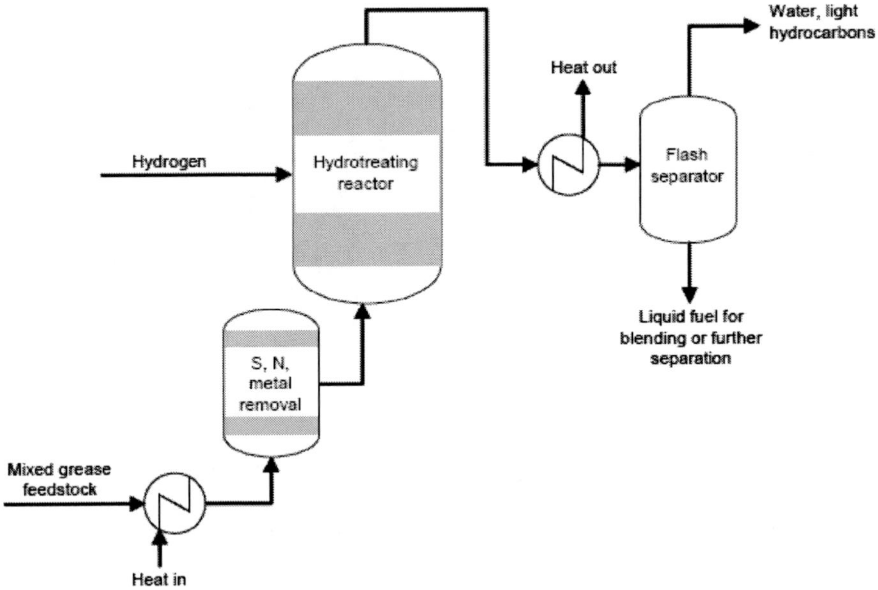

Source: EPA, *Renewable Fuel Standard Program (RFS2) Regulatory Impact Analysis*, EPA420-R-10-006, February 2010.

Figure C-12. Renewable Diesel Production Process.

Dynamic Fuels, LLC, is expected to begin production of renewable diesel in Geismar, LA, over the next few years. The 75 mgpy plant will use Tyson meat processing fats as feedstock to produce diesel fuel using hydrogen supply to be available locally.

Second-Generation (Cellulosic) Biodiesel

Currently, there are no commercial cellulosic diesel plants operating today—most existing plants are pilot and demonstration plants only, producing less than one million gallons annually. However, we project growth in cellulosic diesel plants in 2012 as technologies become commercially viable and cellulosic fuel mandates increase. Until then, pilot plants will likely continue in operation, as technologies to produce diesel from cellulosic biomass are evaluated. In addition, smaller-scale "niche" cellulosic diesel plants may be built to capture favorable economic opportunities.

Production Processes

Cellulosic diesel will likely be produced via two major thermochemical pathways:

- *BTL*. Cellulosic feedstocks are heated to 600°C to 800°C in the presence of a hydrogen source (steam or methane) to form synthetic gas, or "syngas," a mixture of CO and H_2 gases in concentrations tailored to the desired use. The syngas can then be used in an Fischer-Tropsch process or treated with methanol to form diesel products.
- *Pyrolysis*. Cellulosic feedstocks are heated to 500°C to 700°C in the complete absence of oxygen to form an intermediate "bio oil" that can subsequently be processed and/or refined into diesel fuel.

These thermochemical processes generate a "drop-in" diesel fuel (pure hydrocarbon chains without ester links) that can be easily integrated into existing infrastructure. More information on these processes can be found in Chapter 5 under cellulosic ethanol.

Production Locations

Discussion of current and planned locations for cellulosic biofuel production is provided in Chapter 5 under cellulosic ethanol.

Third-Generation (Algal) Biodiesel

Currently, the market for algal-based biofuels is in its infancy. The majority of current production in the United States is relegated to small pilot plants and university and industrial research. One major research initiative started by industry in 2009 is not expected to begin yielding results until 2014 at the earliest.[284] This makes it difficult to forecast the growth of this market.

In 2009, DOE awarded the following grants to support commercialization of biofuels from algal feedstocks, including "Algenol Biofuels ($25 million grant for a pilot scale project located in Freeport, TX), Solazyme ($22 million grant for a pilot scale project located in Riverside, PA) and Sapphire Energy ($50 million grant for a demonstration scale project located in Columbus, NM)."[285]

Many companies have announced plans for commercial scale biofuel production using algal feedstocks, but projections using these announcements are highly uncertain. Planned algae biofuel production before 2020 includes Sapphire Energy (synthetic crude oil to diesel and jet fuel), Petrosun (FAME biodiesel), Solazyme (FAME biodiesel and synthetic crude oil to diesel and jet fuel), and U.S. Biofuels (FAME biodiesel).

Production Processes

There are four main steps involved in producing biofuels from algal feedstocks: a growth stage where the algae to be harvested are allowed to grow in a medium (either fresh or impaired water); a harvest stage where the algae is collected; an extraction stage where the harvested algae is processed into biofuel feedstock; and a production stage were the feedstock is converted into biofuels (typically biodiesel or synthetic crude oil for refining into diesel or jet fuel).

Stage 1: Algae Growth

Algae growth is typically completed either in open pond systems or in closed bioreactors. These two technologies and their respective advantages and disadvantages are discussed later in this document. The algae to be used can be grown in either fresh or impaired waters (saline or wastewater). The algae require a source of energy, typically sunlight but sugar is also being pursued as a potential food source, and CO_2 for photosynthesis, as well as nutrients–nitrogen and phosphorous in particular.

Stage 2: Algae Harvest

There are two main methods of harvesting algal biomass currently used: flocculation and centrifugation.[286] The choice of which harvesting method to use depends significantly on the biomass type and the requirements of downstream processing.

Algae from large volume cultures are typically harvested using centrifugation. This process utilizes high-speed centrifuges to separate the components of the algae so the feedstock can be moved onto the final stage. A major disadvantage to this process is the cost and energy required to do it, both of which are relatively high.

Flocculation is a process by which air bubbles are used to float the desired biomass materials and separate them out. This process has a lower cost than centrifugation, is also more energy efficient and is potentially scalable to larger algal ponds. The downside, though, is that it can yield low concentrations of the final product and may necessitate an additional concentration step.

Stage 3: Feedstock Extraction

Two possible methods of extracting the lipids from the collected algae are drying and cell rupture.[287] The drying process typically utilizes a heat source to evaporate all of the water out of the algae harvested. This process has a major disadvantage in that it is very energy intensive. Utilizing cell rupture techniques (mechanical, chemical, and enzymatic) that have already been applied to oilseeds is another possibility. Utilizing one of these methods, the cell walls of the algae are broken and the different components are then separated. The problem with this technique is that the physical properties of algae strains are different so the technique must be tailored depending on the strain of algae used.

Stage 4: Biofuel Production

Once the oils have been extracted from the algae, a range of conversion pathways can convert the oils into biofuels. FAME biodiesel can be produced via transesterification, similar to first-generation biodiesel. Synthetic crude oil, which can be refined into diesel and jet fuel, can be produced using thermochemical processes, similar to second-generation biodiesel.

Production Techniques

There are two main technologies that can be used to grow algae for biofuel production: open pond systems and closed bioreactors. Open pond systems are typically a collection of "raceways" where the water in each pond is designed as an oval with a paddle wheel being used to circulate the water within the pond. Most closed bioreactor systems are designed as either tubular reactors, plate reactors, or bubble reactors and the system controls water flow and nutrient levels while being exposed to the sunlight of the environment they are housed in.

Open Pond Systems

Figure C-13 shows an example of an open raceway pond system. The majority of algae cultivation done today is done via open raceway pond systems. The main reason for this is the economical nature of these systems. A January 2009 report by the British Columbia Innovation Council found that capital costs of establishing a 400 hectare (roughly 990 acres) open raceway pond system was almost 20 percent of the capital cost of establishing a close bioreactor system of the same size. The same study found that the operational costs of the raceway system were around 50 percent of those of the bioreactor system.[288]

Due to the economics of this system type, open ponds offer the best opportunity for the use of impaired water sources in promoting algae growth. Since the cost of owning and operating an open pond system are significantly less than that of a close bioreactor one, saline water and wastewater can be used, despite the fact that doing so will produce lower quality algal products. This opens the possibility of these types of systems being co-located with wastewater plants throughout the United States, thus enhancing the profit potential of them.

The openness of these systems does present some problems and challenges. Since raceway systems are open to the environment, the possibility of contamination is present. Rainfall and evaporation can change the salinity and the pH of the ponds. Also, these types of systems usually work best with local algae species, which may not yield optimal growth and oil production results. Monocultures can be hard to grow since they are susceptible to local strains. Since these systems typically utilize shallow ponds, topography can also be an issue in development.

Any land that has a slope of more than 5 percent cannot be developed on. The additional costs of developing on this type of land can make it prohibitively expensive.[289]

Source: IEA Greenhouse Gas R&D Program, www.ieaghg.org/.

Figure C-13. Example of an Open Pond Raceway System.

Source: University of Karlsruhe, www.mvm.uni-karlsruhe.de/english/moose.php.

Figure C-14. Example of a Photobioreactor.

Closed Bioreactor Systems

Figure C-14 is an example of a photobioreactor. Being a closed system, bioreactors can answer most of the problems facing open raceway systems. As they are not exposed to the elements, these systems allow for control of the nutrient, pH, salinity, light intensity, and CO_2 levels within the system. Also, any strain of algae can be effectively grown in a bioreactor system, thus allowing for optimal selection of algae to be grown. A bioreactor can also be set up anywhere, thus mitigating the land slope restriction placed upon raceway systems. All of these benefits mean that bioreactor systems can be almost five times as productive as open pond systems with regards to yield volumes.[290]

The most significant issue with bioreactor systems is the cost. As mentioned previously, these systems can cost almost five times as much to build as an open pond system and have double the operating costs. This economic situation leads to other limitations. For one, these costs make only high-value end-products of any worth. This means that impaired water sources, such as wastewater, will most likely not be economical.

Algal Oil Extraction Methods

There are four main methods of extracting oil from the lipids of harvested algae:

- *Chemical extraction.* Solvents, such as benzene, hexane, and ether, can be used to extract oil from harvested algae. The main downside of using chemical extraction methods are the dangers inherent in using the possible solvents—benzene is a carcinogen. There are three methods of chemical extraction of algal oils currently in use: the hexane solvent method, soxhlet extraction, and supercritical fluid extraction.[291]

- *Mechanical extraction.* The simplest form of mechanical oil extraction is using an expression/expeller press and involves crushing. A more sophisticated form of mechanical extraction is ultrasonic-assisted extraction, which uses ultrasonic waves to help break down the cell walls of the algal materials being processed.
- *Enzymatic extraction.* Using enzymes and water as a solvent, the cell walls of the algal material are broken down and the lipids are then transferred to the water. This process makes fractionation of the resulting oil much easier, but currently has the downside that the estimated costs of the process are significantly higher than hexane solvent extraction.[292]
- *Osmotic shock extraction.* By suddenly reducing the osmotic pressure of the algal material, the cells within the material are ruptured, which can release the oil contained within the material enabling its extraction.[293]

Biodiesel Production Costs

First-Generation Biodiesel

FAME–Soybean Oil

Feedstock costs represent most (up to 84 percent) of the total production cost of biodiesel. At the June 2010 price of soybeans ($0.3683 per pound), the net production cost per gallon of ethanol is $3.39.[294] Table C-6 presents a breakdown of the production costs of biodiesel from soybean oil.

In August 2010, the average revenue for first-generation biodiesel was $3.30 per gallon, yielding an overall $0.09 loss per gallon including fixed costs ($0.17 return per gallon not including fixed costs).[295] Biodiesel revenue is maintained by the $1.00 tax credit per gallon, which is crucial to supporting biodiesel profitability.

Renewable Diesel

Similar to the FAME process, feedstock costs represent most (78 percent) of the total production cost of renewable diesel. At a projected price of $0.23 per pound for yellow grease, the net production cost per gallon of renewable diesel is $2.42.[296] Table C-7 presents a breakdown of the production costs of renewable diesel from yellow grease.

Table C-6. Breakdown of Current Net Production Costs for FAME Biodiesel

Production cost component	Production cost ($/gallon)
Corn feedstock	2.76
Natural gas	0.04
Methanol	0.11
Other variable costs	0.25
Fixed costs	0.26
Total production costs	3.42
Sales of byproducts (glycerine)	0.03
Total net production cost	3.39

Table C-7. Breakdown of Projected Production Costs for Renewable Diesel

Production cost component	Production cost ($/gallon)
Yellow grease feedstock	1.90
Hydrogen	0.17
Other variable costs	0.08
Fixed costs	0.27
Total production costs	2.42

Second-Generation Biodiesel

EPA has estimated commercial scale production costs for cellulosic diesel produced from the thermochemical process (Fischer-Tropsch) from wood wastes. Table C-8 shows the breakdown of projected cellulosic diesel costs, which are competitive with projected diesel costs.

Cellulosic diesel production requires higher initial capital investment for biorefinery construction than for first-generation biodiesel. As the technology develops and is refined, these investment requirements should decrease. Average capital costs for a 33.2 million gallon capacity cellulosic diesel plant are projected to be $346 million.[297] These costs are three to four times the capital investment costs for non-cellulosic alternatives.

Third-Generation (Algal) Biodiesel

EPA has estimated commercial scale production costs for FAME biodiesel produced from algal feedstocks. Algae oil feedstock costs are projected to reach $0.58 per pound, supporting a fuel production cost of $4.52 per gallon. However, estimates of these costs are highly variable, with various literature sources forecasting costs between $2.50 and $20.00 per gallon.

Biodiesel Tax Credit and Implications

In 2004, the American Jobs Creation Act of 2004 (H.R. 4520, Public Law 108-357) established biodiesel producer tax credits of $1.00 per gallon for "agribiodiesel" (biodiesel produced from agricultural products such as soybean oil or animal fats), or $0.50 per gallon for biodiesel produced from other sources (e.g., recycled vegetable oil). These tax credits were extended and expanded by the Emergency Economic Stabilization Act of 2008, which also increased the credit to $1.00 for all biodiesel and applied to biodiesel made from recycled vegetable oils or animal fats, as well as renewable diesel (e.g., diesel derived from biomass). Although this tax credit expired on December 31, 2009, it is likely that the Congress will reinstate the credit during 2010.

First-Generation Biodiesel Production Tax Credits

With a tax credit of $1.00 per gallon, first-generation ethanol production costs are reduced to $2.39, and biodiesel remains somewhat competitive with diesel. Note that the tax credit actually is provided directly to biodiesel blenders, and it is implied in the comparison of production costs with diesel. As shown in Figure C-15, the production cost of biodiesel rises and falls with the price of soybean oil. The current implied biodiesel production price of $2.39 per gallon (with tax credit) is equivalent to diesel production with crude oil at roughly $95 per barrel. Without the tax credit, the equivalent crude oil price jumps to $136 per barrel.

Table C-8. Breakdown of Projected Biochemical Cellulosic Diesel Production Costs (Wood Wastes)

Production cost component	Production cost ($/gallon)	Percent
Biomass	1.17−1.56	40−47
Feed handling	Included above	
Other raw materials	0.05	2
Waste disposal and catalyst	0.05	2
Capital costs	1.14	34−39
Fixed costs	0.54	16−18
Total production costs	2.95−3.34	100
Naphtha and wax revenue	0.97	
Total net production cost	1.98−2.37	

Source: EPA, *Renewable Fuel Standard Program (RFS2) Regulatory Impact Analysis*, EPA-420-R-10-006, February 2010.

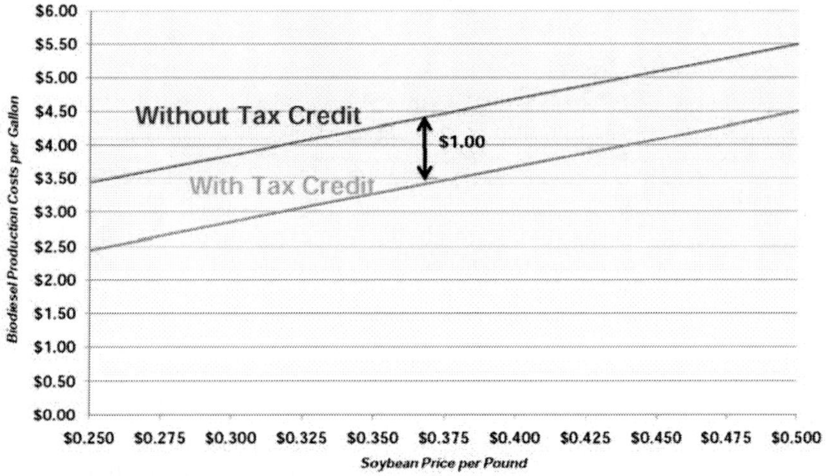

Figure C-15. Net Biodiesel Production Costs and Commodity Price of Soybean Oil.

The Small Agri-Biodiesel Producer Credit provides an additional tax credit for small biodiesel production (annual capacity less than 60 million gallons) of $0.10 per gallon on the first 15 million gallons of ethanol produced in a tax year. This credit also expired in December 2009.

Second-Generation Cellulosic Diesel Production Tax Credits

In May 2008, the Food, Conservation, and Energy Act of 2008 (Public Law 110234, Section 15321) established the CBPTC. The CBPTC provides a tax credit of up to $1.01 for producers of cellulosic diesel. The CBPTC is set to expire on December 31, 2012. This tax credit is intended to support the early growth of the cellulosic biofuels industry.

Future of Biodiesel, Renewable, and Cellulosic Diesel Tax Credits

In May 2010, H.R. 4213, *The American Jobs and Closing Tax Loopholes Act of 2010*, was introduced, which would extend the $1.00 per gallon production tax credit for biodiesel and the small agri-biodiesel producer credit of 10 cents per gallon through December 31, 2010. In July 2010, the U.S. House Committee on Ways and Means introduced *The Domestic, Manufacturing and Energy Jobs Act of 2010*, which would reinstate the biodiesel tax incentives for 2011. If these credits are not extended, the viability of the domestic biodiesel industry will be in question.

Biodiesel Transportation

FAME Biodiesel

Transportation of FAME biodiesel encounters issues similar to ethanol—rail tank cars and tanker trucks have limited availability, and a pipeline transportation alternative is lacking. The lower production and distribution volumes for biodiesel compared with ethanol compound the transportation issues in that diseconomies of scale are accentuated. However, biodiesel production and point of sale are generally more dispersed than for ethanol, which could help support faster development of transportation infrastructure.

Renewable and Cellulosic Diesel

The key advantage of renewable and cellulosic diesel compared to FAME biodiesel is that the diesel fuel produced is fungible within the existing petroleum supply system, and can use the existing transportation supply infrastructure. Therefore, these fuels do not carry the same transportation and other supply chain issues as FAME biodiesel. Renewable and cellulosic diesel can be transported through the same pipelines, use the same refueling infrastructure, and be used in the same vehicles as diesel fuel.

FAME Biodiesel Blending

FAME biodiesel distribution is complicated by storage challenges, blending limitations, and physical state issues. (These issues do not apply to renewable and cellulosic diesel). Pure biodiesel (B100) is believed to degrade to below acceptable quality if stored for periods of more than 6 months.[298] Biodiesel can also dissolve accumulated sediments in storage and engine fuel tanks, so tanks must be cleaned thoroughly before biodiesel is added. These dissolved sediments could cause eventual fuel injection failure.[299] The NBB recommends "B100 be shipped in a way that does not lead to contamination. The association says trucks and/or railcars should be washed out before being loaded—and the only residual that is acceptable in a tanker is petroleum diesel."[300]

The EIA handling guide reports,

> As demand for biodiesel increases, petroleum terminals and pipeline racks are installing biodiesel blending capability so that jobbers and distributors can receive a biodiesel blend directly at the rack and store and distribute only the blended biodiesel. This finished blend can then be sold to fleet or other applications that have some type of on-site storage. Even more recently, there are an increasing number of public pumps and key card pumps that are carrying biodiesel blends for individual users or for fleets who do no t have their own on-site storage capability. As the market matures and volumes

continue to increase, it is likely that the actual point of blending will occur further and further upstream in the distribution system.[301]

Biodiesel can be blended with petroleum diesel at any concentration to produce a biodiesel blend, but there are specific blends necessary to meet existing ASTM fuel specifications. At blends of 5 percent and lower (i.e., B2 and B5), the fuel can be treated as fungible, conventional diesel per ASTM D975. Biodiesel blends between 6 percent and 20 percent fall under the ASTM standard D7467 and are marketed as commercial biodiesel.

Biodiesel can be blended by one of following primary processes:

- B100 (100 percent biodiesel) splash blended with diesel fuel by the end user
- Blended by a jobber or distribution company and offered for sale as a finished blend
- Blended at a petroleum terminal or rack by a pipeline or terminal company and offered as a finished blend. This product is sold directly to customers or to a petroleum jobber or distribution company for further sale to customers.[302]

Biodiesel Retail Sales

FAME Biodiesel

Although biodiesel retail sales have grown significantly in the past two decades, consumer concerns over the image of diesel as a dirty fuel, biodiesel's performance issues in cold weather, and uncertainty over biodiesel's impact on engine warranties have hindered growth. ULSD regulations passed in October 2006 may help to combat the negative image of diesel, and biodiesel's lubricity characteristics may encourage the use of the fuel as an additive to correct ULSD's low lubricity.

We forecast production and sales of FAME biodiesel on the basis of FAPRI data. As shown in Table C-9 and Figure C-16, we project domestic FAME biodiesel production and sales to increase nearly threefold between 2009 and 2020, from 475 to 1,418 million gallons. Most of this biodiesel will be blended and sold as B20.

Table C-9. Projected FAME Biodiesel Production and Retail Sales, 2009–20 (million gallons)

	2009	2010	2011	2012	2013	2014	2015	2016	2017	2018	2019	2020
Vegetable oils	434	598	724	846	914	914	956	1,037	1,081	1,107	1,120	1,108
Fats and greases	41	46	63	79	96	118	130	150	170	190	210	230
Algae	0	0	0	0	0	0	0	10	20	40	60	80
Total FAME biodiesel production	475	644	787	925	1,010	1,032	1,086	1,197	1,271	1,337	1,390	1,418

Note: Fuel quantities represent "neat" fuel rather than blended fuel.

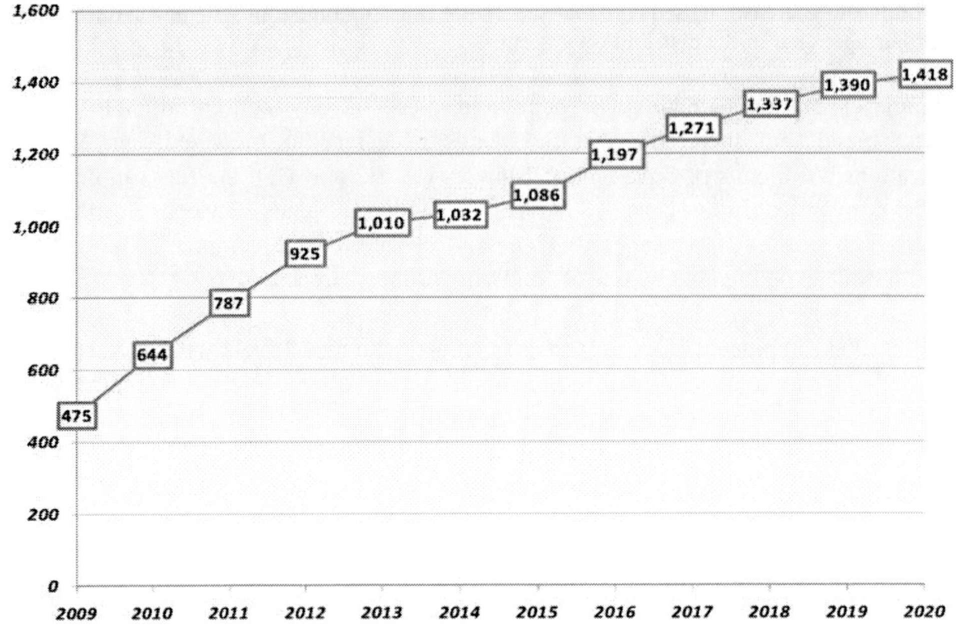

Figure C-16. Projected FAME Biodiesel Production and Retail Sales, 2009–20 (million gallons).

DOE estimates that B20 is available today at only around 652 (or 0.4 percent) of the roughly 170,000 fuel stations in the United States.[303] Unlike E85 stations, which are concentrated near production facilities, B20 stations are more dispersed throughout the United States (Figure C-17).

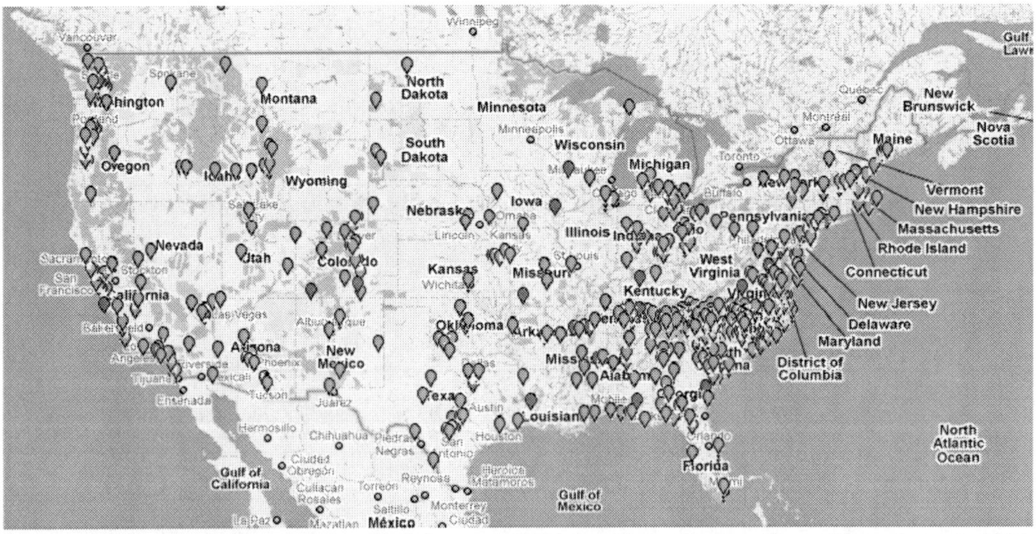

Source: DOE, Alternative Fuels and Advanced Vehicles Data Center, *TransAtlas*, August 2010, rpm.nrel.gov/transatlas/launch.

Figure C-17. Location of Commercial B20 Stations, August 2010.

Biodiesel Warranty Issues

Some vehicle owners and operators express concern over the warranty impacts of using biodiesel in their diesel vehicles. Most engine and vehicle manufacturers will not cover damage caused by an external condition, such as the quality of fuel used in the vehicle. However, the NBB explains, "If an engine that uses biodiesel experiences a failure unrelated to biodiesel use, it must be covered by the Original Equipment Manufacturer's warranty. Federal law (The Magnuson Moss Act), prohibits the voiding of a warranty just because biodiesel was used—it has to be the cause of the failure."

Biodiesel Retail Prices

B20 is competitively priced compared with diesel. Unlike ethanol, B20 has a comparably high energy content to diesel—98.2 percent of the energy content. In April 2010, the average price of B20 at the pump was $3.12 per gallon (or $2.85 per GGE).[304] This represents a 3 percent premium over the per gallon price of diesel ($3.02) and 5 percent premium over the per GGE diesel price ($2.71).[305]

Renewable Diesel

Since renewable diesel fuel is fungible with the existing diesel supply, the only limitation of sales will likely be cost-effective production volumes. Using EPA data, we forecast renewable diesel production. As shown in Table C-10 and Figure C-18, we project domestic renewable diesel production to grow to 150 million gallons by 2020.

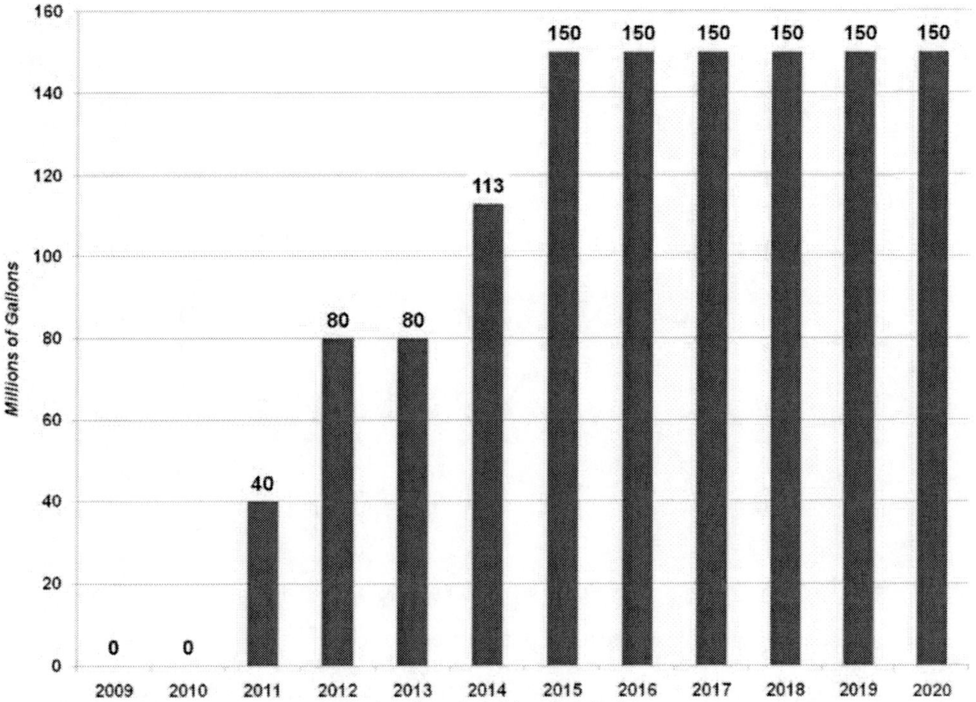

Note: Fuel quantities represent "neat" fuel rather than blended fuel.

Figure C-18. Projected Renewable Diesel Production and Retail Sales, 2009–20 (million gallons).

Table C-10. Projected Renewable Diesel Production and Retail Sales, 2009–20 (million gallons)

2009	2010	2011	2012	2013	2014	2015	2016	2017	2018	2019	2020
0	0	40	80	80	113	150	150	150	150	150	150

Cellulosic Diesel

Similar to renewable diesel fuel, cellulosic diesel fuel is fungible with existing diesel supply. Therefore, the only limitation of sales will likely be cost-effective production volumes. Again using EPA data, we forecast cellulosic diesel production. As shown in Table C-11 and Figure C-19, we project domestic cellulosic diesel production to grow to 4,280 million gallons by 2020.

Table C-11. Projected Cellulosic Diesel Production and Retail Sales, 2009–20 (million gallons)

2009	2010	2011	2012	2013	2014	2015	2016	2017	2018	2019	2020
0	0	0	100	410	710	1,220	1,730	2,240	2,850	3,460	4,280

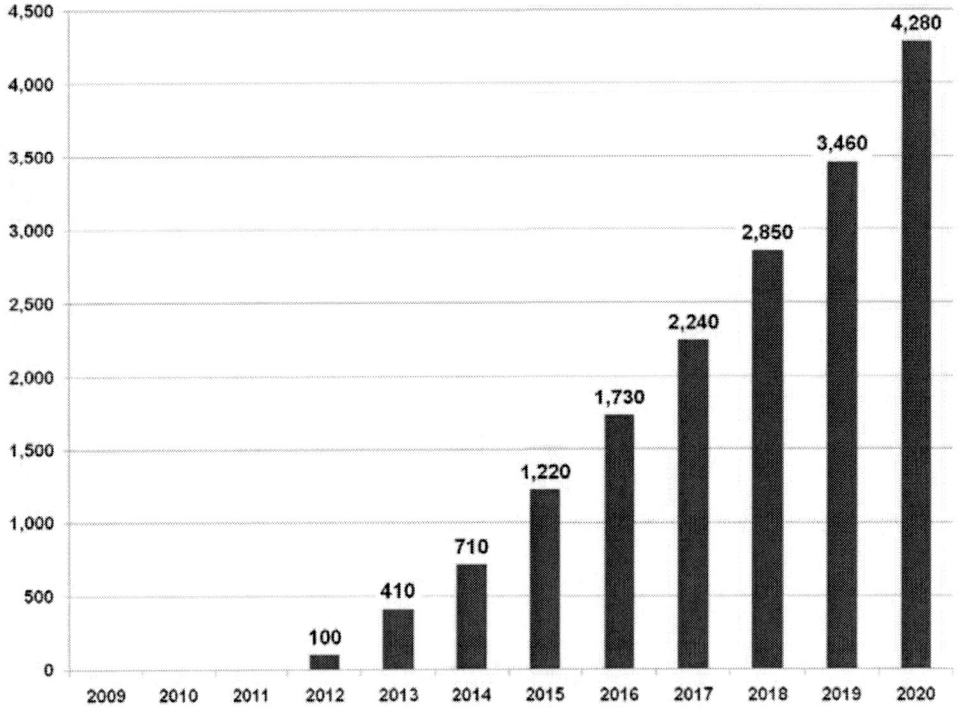

Note: Fuel quantities represent "neat" fuel rather than blended fuel.

Figure C-19. Projected Cellulosic Diesel Production and Retail Sales, 2009–20 (million gallons).

APPENDIX D. BIOBUTANOL MARKET SUMMARY

In this appendix, we present our assessment of the current and future commercial availability of biobutanol. We discuss the production of biobutanol through FY20 and identify the primary factors that will limit the commercial availability of this biofuel.

Butanol is a four-carbon alcohol (butyl alcohol) produced today almost entirely from petroleum. Its primary use is as an industrial solvent in products such as lacquers and enamels.

Compared to bioethanol, biobutanol is considered a next-generation biofuel. Very little biobutanol is in production today. However, it has a number of advantages over ethanol as a renewable alternative fuel, including the following:

- Energy content closer to that of gasoline than ethanol, which is important as the amount of biofuel in the fuel blend increases.
- Suitability to current vehicle and engine technologies. For example, ButylFuel, LLC, drove an unmodified 1992 Buick Park Avenue car across the country in 2005 fueled with 100 percent biobutanol at a fuel economy of 24 miles per gallon (mpg) versus 22 mpg for the same model fueled with gasoline.
- No requirement for flex-fuel vehicle pipes and hoses.
- Does not require automakers to compromise on performance to meet environmental regulations.

Fuel and fleet testing of biobutanol in existing vehicles on roads—covering a distance of more than 1.3 million vehicle road-miles—has suggested the high performance advantages of biobutanol. A commercial fuels trial confirmed the compatibility of butanol with existing fuel infrastructure and consumer satisfaction with the product.[306]

Biobutanol Supply Chain Overview

As a result of its properties, biobutanol is receiving increased attention as a means to achieve greater energy independence and energy security, and to reduce climate change and economic growth concerns. The major hurdle to using this fuel is commercial production and availability, which at this time are exceptionally limited and highly regionalized.

Biobutanol Feedstocks

Since the early 1900s it has been possible to produce butanol from biomass sources via the acetone-butanol-ethanol fermentation process using Clostridium acetobutylicum. However, since the 1950s, these older butanol production processes have been more expensive than petrochemical production of butanol.

Renewed interest in butanol as a sustainable vehicle fuel has resulted in the development of improved biobutanol production processes. These improved processes may use traditional feedstocks, such as sugar beets, corn, sugarcane, animal fats, and vegetable oils. Recently

ButylFuel, LLC, used a DOE Small Business Technology Transfer grant to develop a process aimed at making biobutanol production economically competitive with petrochemical production processes.[307]

Biobutanol Production and Transportation

The USDA Economic Research Service estimated that 2010 biobutanol production capacity in the United States was only 1.0 mpgy. It further projected that Gevo would add 50 mpgy of biobutanol production capacity by 2011.[308] Unlike ethanol, biobutanol can be transported through gasoline infrastructure and pipelines.

Blending

Biobutanol has the benefit of blending at higher concentrations than bioethanol for use in standard vehicle engines. Currently, it can be blended up to 10 percent volume over volume (v/v) in European gasoline and 11.5 percent v/v in U.S. gasoline. The potential exists to increase the maximum allowable use in gasoline up to 16 percent volume.

Biobutanol is also less susceptible to blend separation in the presence of water than ethanol and gasoline blends. This characteristic allows its use within the energy industry's existing distribution infrastructure without requiring modifications in blending facilities, storage tanks, or retail station pumps. EPA regulations permit blending of biobutanol as an oxygenate with gasoline in concentrations up to 11.5 percent by volume. Blends of 85 percent or more biobutanol with gasoline are required to qualify as an EPAct alternative fuel.

Biobutanol advocates claim that today's vehicles can be fueled with high concentrations of biobutanol (up to 100 percent) with minor or no vehicle modifications, although testing of this claim has been limited.[309]

Biobutanol Costs

Based upon information from various commercial biobutanol production ventures, production costs may decline if companies can realize cost efficiencies from their technological advances in strain improvement, reaction management, and separation technology to make biobutanol production a sustainable commercial enterprise. By designing for a diversity of feedstocks, the production facilities could accept a range of regional, low-cost plant materials to manage feedstock expenses. The industry is currently just beginning to project possible commercial scale costs.

Biobutanol Retail Sales

Biobutanol is not readily available for purchase in commercial scale quantities because of the similar market niche currently dominated by ethanol. In the absence of a specific mandate or shift in market demand, this fuel's sales are not anticipated to expand greatly by 2020. However, recent development in biobutanol conversion to diesel and jet fuels via the oligomerization pathway, developed by the U.S. Naval Air Warfare Center at China Lake, CA, could significantly increase commercial viability and sales of biobutanol as an intermediate product.

APPENDIX E. DEFINITIONS

Document Definitions

Alternative fuels. Transportation or mobility fuels, including renewable and synthetic fuels, not composed of or derived from liquid petroleum. These fuels can include petroleum liquid and alcohol blends containing 15 percent or less of petroleum that are pursuant to standard seasonal fuel specifications.

B2, B5, B20, and B100. B2 is 2 percent biodiesel and 98 percent diesel; B5 is 5 percent biodiesel; B20 is 20 percent biodiesel; and B100 is pure biodiesel. Most biodiesel is sold as B20.

Biobutanol. Butanol from biomass is called biobutanol.

Bio-derived synthetic paraffinic kerosene (bio-SPK). Oil created from camelina seeds that can be processed into a third-generation, drop-in replacement for petroleum-based jet fuels, known as HRJ or bio-SPK.[310]

Camelina. Native to the Mediterranean and Central Asia, *Camelina sativa,* gold of pleasure, linseed dodder, or false flax is an oil-seed plant that belongs to the mustard family.

Cellulosic biodiesel. Biodiesel produced from cellulosic biomass using thermochemical processes, including the Fischer-Tropsch BTL and other BTL processes.

Cellulosic biomass. Corn stover, timber wastes, and dedicated energy crops such as switchgrass.

Class III bulk petroleum. Of the U.S. armed forces classes of supply, the third class, Class III, comprises POL, package and bulk, including petroleum, fuels, lubricants, hydraulic and insulating oils, preservatives, liquids and gases, bulk chemical products, coolants, deicer and antifreeze compounds, components, additives of petroleum and chemical products, and coal.

E85 and E15. Ethanol fuel mixtures have "E" numbers, which describe the percentage of ethanol in the mixture by volume. E85 is 85 percent anhydrous ethanol and 15 percent gasoline. Low ethanol blends, from E5 to E25, are also known as *gasohol*, though internationally the term most commonly refers to the E10 blend.

Ethanol. Ethyl alcohol, produced through the fermentation and distillation of simple sugars.

Fatty acid methyl ester (FAME)–based biodiesel. Most biodiesel produced today if from FAME processes, which use vegetable oils or animal fats to produce biodiesel through transesterification.

Fischer-Tropsch (FT) process. A set of chemical reactions that convert a mixture of carbon monoxide and hydrogen into liquid hydrocarbons. The process, a key component of gas-to-liquids technology, produces a petroleum substitute, typically from coal, natural gas, or biomass for use as synthetic lubrication oil and as synthetic fuel.

Fungibility. The ease of exchanging one unit of a commodity with another unit of the same commodity.

Gasohol. Corn-based ethanol, typically a blend of 90 percent gasoline and 10 percent ethanol, also referred to as E10.

Hydrotreated renewable jet (HRJ) fuel. A drop-in aviation fuel that can be produced from triglyceride feedstocks, such as animal fats and greases (such as tallow); plant oils (such as camelina, jatropha, soy, and canola); and algal oils.

Renewable biomass. Planted crops and crop residue harvested from existing agricultural land cleared or cultivated prior to December 19, 2007, which was nonforested and either actively managed or fallow on December 19, 2007; planted trees and tree residue from a tree plantation located on non-federal land that was cleared and actively managed prior to December 19, 2007; animal waste material and animal byproducts; slash and pre-commercial thinnings from non-federal forestland that is not ecologically sensitive; biomass (organic matter available on a renewable or recurring basis) obtained from the immediate vicinity of buildings and other areas regularly occupied by people, or of public infrastructure, in an area at risk of wildfire; algae; separated yard waste or food waste, including recycled cooking and trap grease.

Renewable diesel. Diesel fuel derived from biomass using a thermal depolymerization process (currently used in petroleum refining) that meets the requirements of ASTM D975 for petroleum diesel fuel.

Renewable fuels. Transportation or mobility fuels, used alone or blended with petroleum-based fuel, and wholly derived from biomass or its decay products. (Also, petroleum-blended fuel with a renewable component above a certain percentage of "neat" renewable fuel products, for example, B20 is termed biodiesel.)

Synthetic fuels. Liquid hydrocarbon fuels produced from coal, natural gas, or, increasingly, biomass.

Synthetic paraffinic kerosene (SPK). HRJ feedstocks are converted through hydrotreating to produce renewable SPK jet fuel, which is differentiated from cellulosic Fischer-Tropsch process-derived SPK.

Transesterification. The process of exchanging the organic group R "of an ester with the organic group R" of an alcohol. These reactions are often catalyzed by the addition of an acid or base catalyst. The reaction can also be accomplished with the help of enzymes (biocatalyst), particularly lipases.

Definitions from EISA 2007 and RFS2

Algae includes cyanobacteria and diatoms, but not microcrop angiosperms.

Areas at risk of wildfire are those areas in the "wildland-urban interface," where humans and their development meet or intermix with wildland fuel. Note that, for guidance, the SILVIS laboratory at the University of Wisconsin maintains a website that provides a detailed map of areas meeting this criteria at: www.silvis.forest.wisc.edu/projects/US_WUI_2000.asp. The SILVIS laboratory is located at 1630 Linden Drive, Madison, Wisconsin 53706 and can be contacted at 608-263-4349.

Biomass is organic matter that is available on a renewable or recurring basis.

Crop residue is the biomass left over from the harvesting or processing of planted crops from existing agricultural land and any biomass removed from existing agricultural land that facilitates crop management (including biomass removed from such lands in relation to invasive species control or fire management), whether or not the biomass includes any portion of a crop or crop plant.

Cropland is land used for production of crops for harvest and includes cultivated cropland, such as for row crops or close-grown crops, and non-cultivated cropland, such as for horticultural or aquatic crops.

Ecologically sensitive forestland means forestland that meets either of the following criteria:

1. An ecological community with a global or state ranking of critically imperiled, imperiled or rare pursuant to a State Natural Heritage Program. For examples of such ecological communities, see "Listing of Forest Ecological Communities Pursuant to 40 CFR 80.1401; S1–S3 communities," which is number EPA–HQ–OAR–2005–0161–1034.1 in the public docket, and "Listing of Forest Ecological Communities Pursuant to 40 CFR 80.1401; G1–G2 communities," which is number EPA–HQ–OAR–2005– 0161–2906.1 in the public docket. This material is available for inspection at the EPA Docket Center, EPA/DC, EPA West, Room 3334, 1301 Constitution Avenue, NW, Washington, DC. The telephone number for the Air Docket is 202-566-1742.

2. Old growth or late successional, characterized by trees at least 200 years in age.

Existing agricultural land is cropland, pastureland, and land enrolled in the Conservation Reserve Program (administered by USDA's Farm Service Agency) that was cleared or cultivated prior to December 19, 2007, and that, on December 19, 2007, was:

1. Nonforested; and
2. Actively managed as agricultural land or fallow, as evidenced by records which must be traceable to the land in question, which must include one of the following:
   a. Records of sales of planted crops, crop residue, or livestock, or records of purchases for land treatments such as fertilizer, weed control, or seeding.
   b. A written management plan for agricultural purposes.
   c. Documented participation in an agricultural management program administered by a federal, state, or local government agency.
   d. Documented management in accordance with a certification program for agricultural products.

Fallow means cropland, pastureland, or land enrolled in the Conservation Reserve Program (administered by USDA's Farm Service Agency) that is intentionally left idle to regenerate for future agricultural purposes with no seeding or planting, harvesting, mowing, or treatment during the fallow period.

Forestland is generally undeveloped land covering a minimum area of 1 acre upon which the primary vegetative species are trees, including land that formerly had such tree cover and that will be regenerated and tree plantations. Treecovered areas in intensive agricultural crop production settings, such as fruit orchards, or tree-covered areas in urban settings, such as city parks, are not considered forestland.

Nonforested land means land that is not forestland.

Pastureland is land managed for the production of select indigenous or introduced forage plants for livestock grazing or hay production, and to prevent succession to other plant types.

Planted crops are all annual or perennial agricultural crops from existing agricultural land that may be used as feedstocks for renewable fuel, such as grains, oilseeds, sugarcane, switchgrass, prairie grass, duckweed, and other species (but not including algae species or planted trees), providing that they were intentionally applied by humans to the ground, a growth medium, a pond or tank, either by direct application as seed or plant, or through intentional natural seeding or vegetative propagation by mature plants introduced or left undisturbed for that purpose.

Planted trees are trees harvested from a tree plantation.

Pre-commercial thinnings are trees, including unhealthy or diseased trees, removed to reduce stocking to concentrate growth on more desirable, healthy trees, or other vegetative material that is removed to promote tree growth.

Slash is the residue, including treetops, branches, and bark, left on the ground after logging or accumulating as a result of a storm, fire, delimbing, or other similar disturbance.

Tree plantation is a stand of no less than 1 acre composed primarily of trees established by handor machine-planting of a seed or sapling, or by coppice growth from the stump or root of a tree that was handor machine-planted. Tree plantations must have been cleared prior to December 19, 2007 and must have been actively managed on December 19, 2007, as evidenced by records which must be traceable to the land in question, which must include

1. sales records for planted trees or tree residue together with other written documentation connecting the land in question to these purchases;
2. purchasing records for seeds, seedlings, or other nursery stock together with other written documentation connecting the land in question to these purchases;
3. a written management plan for silvicultural purposes;
4. documentation of participation in a silvicultural program sponsored by a federal, state, or local government agency;
5. documentation of land management in accordance with an agricultural or silvicultural product certification program;
6. an agreement for land management consultation with a professional fore-ster that identifies the land in question; or
7. evidence of the existence and ongoing maintenance of a road system or other physical infrastructure designed and maintained for logging use, to-gether with one of the above-mentioned documents.

Tree residue is slash and any woody residue generated during the processing of planted trees from tree plantations for use in lumber, paper, furniture, or other applications, provided that such woody residue is not mixed with similar residue from trees that do not originate in tree plantations.

Yard waste is leaves, sticks, pine needles, grass and hedge clippings, and similar waste from residential, commercial, or industrial areas (but not from forestlands or tree plantations).[311]

APPENDIX F. ABBREVIATIONS

AEO	*Annual Energy Outlook*
AER	Annual Energy Review
AESIS	*Army Energy Security Implementation Strategy*
AFB	Air Force Base
AFCO	Alternative Fuels Certification Office
AFMC	Air Force Materiel Command
AFPA	Air Force Petroleum Agency
AFRL	Air Force Research Laboratory
AFV	alternative fuel vehicle
AMRDEC	Aviation and Missile Research Development and Engineering Center

ARNL	Argonne National Laboratory
ASD (OEPP)	Assistant Secretary of Defense for Operational Energy Plans and Programs
ASTM	American Society for Testing and Materials
ATA	Air Transport Association
b/d	barrels per day
BSM-E	Business System Modernization-Energy
BTL	biomass-to-liquids
C4ISR	command, control, communications, computers, intelligence, surveillance and reconnaissance
CAA	Clean Air Act
CBO	Congressional Budget Office
CBOT	Chicago Board of Trade
CBPTC	Cellulosic Biofuel Producer Tax Credit
CCS	carbon capture and storage
CERDEC	Communications-Electronics Research, Development, and Engineering Center
CFR	Code of Federal Regulations
CPI	consumer price index
CWA	Clean Water Act
DARPA	Defense Advanced Research Projects Agency
DDGS	dry distillers grains with solubles
DFSP	defense fuel support point
DGS	distillers' grains with solubles
DLA	Defense Logistics Agency
DoD	Department of Defense
DOE	Department of Energy
DOEPP	Director of Operational Energy Plans and Programs
DPA	Defense Production Act
E^2O	Expeditionary Energy Office
EIA	Energy Information Administration
EISA	Energy Independence and Security Act
EO	Executive Order
EPA	Environmental Protection Agency
EPAct	Energy Policy Act
ETS	Emissions Trading System
EU	European Union
FAME	fatty acid methyl ester
FAPRI	Food and Agricultural Policy Research Institute
FAST	Federal Automotive Statistical Tool
FFV	flex-fuel vehicles
FT-SPK	Fischer-Tropsch synthetic paraffinic kerosene
FY	fiscal year
GAO	Government Accountability Office
GGE	gasoline gallon equivalents
GHG	greenhouse gas

HRJ	Hydrotreated renewable jet fuel
mgpy	million gallons per year
MIT	Massachusetts Institute of Technology
MOU	memorandum of understanding
MSW	municipal solid wastes
MTBE	methyl tertiary butyl ether
NAVSUP	Naval Supply and Systems Command
NBB	National Biodiesel Board
NDAA	National Defense Authorization Act
NETL	National Energy Technology Laboratory
NEV	Neighborhood Electric Vehicle
NPDES	National Pollutant Discharge Elimination System
NSN	national stock number
NTV	non-tactical vehicle
OASD (OEPP)	Office of the Assistant Secretary of Defense for Operational Energy Plans and Programs
ODOEPP	Office of the Director of Operational Energy Plans and Programs
OSD	Office of the Secretary of Defense
PACOM	U.S. Pacific Command
PM-MEP	Program Manager-Mobile Electric Power
POL	petroleum, oil, and lubricants
R&D	research and development
RDECOM	Research, Development and Engineering Command RDT&E research, development, test, and evaluation
RFI	request for information
RFS	Renewable Fuel Standard
RIN	renewable identification number
SEPTC	Small Ethanol Producer Tax Credit
SIP	State Implementation Plan
TARDEC	Tank and Automotive Research, Development and Engineering Center
UCS	Union of Concerned Scientists
UL	Underwriters Laboratories
ULSD	ultra-low sulfur diesel
USDA	U.S. Department of Agriculture
USMC	U.S. Marine Corps
v/v	volume over volume
VEETC	Volumetric Ethanol Excise Tax Credit

End Notes

[1] DoD Quadrennial Defense Review, 2010.

[2] Defense Energy Support Center, Factbook FY09, www.desc.dla.mil/DCM/Files/ FY09%20Fact%20Book%20%288-10-10%29.pdf.

[3] EIA Annual Energy Review (AER), Table 1.13, www.eia.doe.gov/emeu/aer/pdf/pages/sec1_29.pdf.

[4] EIA AER Table 1.3, www.eia.doe.gov/emeu/aer/pdf/pages/sec1_9.pdf.

[5] EIA AER Table 5.14c, www.eia.doe.gov/emeu/aer/pdf/pages/sec5_37.pdf.

[6] Includes JP-4, JAB, JAA, JA1, JP-5, JP8, JPTS, and distillates and diesel, DESC Factbook 2009.

[7] However, most of the statutes and regulations focus on NTV and fleet-applicable fuels rather than those destined for use in tactical systems and weapons platforms.

[8] 40 CFR 80.1101, from P.L. 109-58.

[9] 40 CFR 80.1401, from P.L. 110-140.

[10] 40 CFR 80.1401.

[11] EPA, *EPA Finalizes Regulations for the National Renewable Fuel Standard Program for 2010 and Beyond*, Report EPA-420-F-10-007, February 2010.

[12] EISA, Public Law 110-140, December 19, 2007.

[13] EISA, Sec. 201 (1)(C).

[14] The proposed RFS2 Rule used Argonne National Laboratory's GREET model (Version 1.8b) to determine this baseline. This model uses "average values as inputs to estimate aggregate emissions, rather than using site-specific values" (RFS2, Preamble V(B)(3)). Public comments identified several weaknesses of using the GREET tool for this purpose, including incorrect input values for energy efficiency values for crude oil extraction, methane emission factors for oil pro- duction and flaring, transportation distances for crude oil and petroleum products, and oil tanker payload value; lack of accounting for energy consumption associated with crude oil transport in the country of extraction; and inconsistency in definition of low-sulfur diesel with EPA definition (RFS2, Preamble V(B)(3)).

[15] DOE, NETL, *Development of Baseline Data and Analysis of Life Cycle Greenhouse Gas Emissions of Petroleum-Based Fuel,* 2008, p. ES-1, www.netl.doe.gov/energy-analyses/pubs/NETL%20LCA%20Petroleum-Based%20Fuels%20Nov%202008.pdf.

[16] Regulations.gov Docket ID: EPA-HQ-OAR-2005-0161-3144, www.regulations.gov/search/Regs/home.html#documentDetail?R=0900006480a862a2.

[17] AFRL, *Propulsion and Power Rapid Response Research and Development (R&D) Support, Framework and Guidance for Estimating Greenhouse Gas Footprints of Aviation Fuels*, AFRL- RZ-WP-TR-2009-2206, 2009.

[18] EPA, Regulatory Announcement: EPA Lifecycle Analysis of Greenhouse Gas Emissions from Renewable Fuels, 2010, www.epa.gov/otaq/renewablefuels/420f10006.pdf.

[19] See footnote 11, this chapter.

[20] Robert H. Williams, Eric D. Larson, and Haiming Jin, "Synthetic fuels in a world with high oil and carbon prices," Table 1, prepared for the 8th International Conference on Greenhouse Gas Control Technologies, Trondheim, Norway, June 19–22, 2006, (estimating that CTL synfuel pro- duces 1.8 times the GHG emissions of conventional fuel).

[21] Using GREET 1.8b (March 17, 2008); EPA, *Greenhouse Gas Impacts of Expanded Renew- able and Alternative Fuels Use*, EPA420-F-07-035, fact sheet, April 2007, www.epa.gov/oms/renewablefuels/420f07035.htm).

[22] P.L. 109-58.

[23] P.L. 110-140.

[24] 72 FR 23900.

[25] 72 FR 23900.

[26] See footnote 5, this chapter.

[27] For the purpose of this report, statutes and regulations imply legally binding requirements, while policies and goals imply self-imposed activities.

[28] RAND and Massachusetts Institute of Technology (MIT), *Near Term Feasibility of Alter- native Jet Fuels*, Technical Report, 2009.

[29] EPA, Guidance on State Implementation Plan (SIP) Credits for Emission Reductions from Electric-Sector Energy Efficiency and Renewable Energy Measures, August 2004.

[30] www.epa.gov/climatechange/initiatives/index.html.

[31] President Obama Sets Greenhouse Gas Emissions Reduction Target for Federal Operations, January 29, 2010, www.whitehouse.gov/the-press-office/president-obama-sets-greenhouse-gas-emissions-reduction-target-federal-operations.

[32] See footnote 22, this chapter.

[33] The RAND and MIT (2009) study used the GREET method to estimate GHG emissions.

[34] ARNL, Consumptive Water Use in the Production of Ethanol and Petroleum Gasoline, ANL/ESD/09-1, 2009.

[35] See footnote 28, this chapter.

[36] Winrock International, The Role of Water in the Sustainable Supply of Biofuels, 2009.

[37] ARNL used consumptive water use, which includes irrigation water incorporated into the crop or lost to evapotranspiration. It does not include water output that is recycled and reused.

[38] See footnote 28, this chapter.

[39] DoD, "Office of the Secretary of Defense," www.defense.gov/osd/; DoD, "Department of Defense Directives, Instructions and Regulations for Transportation Policy," www.acq.osd.mil/log/tp/tp_library.htm, and DoD; and Office of the Deputy Under Secretary of Defense for Logistics and Materiel Readiness, 4500.36-R, *Management, Acquisition, and Use of Motor Vehicles*, March 16, 2007, www.acq.osd.mil/log/tp/450036r_1.pdf.

[40] Volume I, Chapter 4, "Fuel Standardization and Cataloging."
[41] Volume II, Defense Fuel Support (DFSP) Management Training Policy.
[42] Volume III, Chapter 8, "DoD Bio-Fuels Program Biodiesel Blend (B20) and Fuel Ethanol Blend (ED 75-ED 85)."
[43] This applies to DoD and federal locations in the domestic United States, including Hawaii and Alaska.
[44] Section 8b of EO 13423 "Strengthening Federal Environmental Energy, and Transportation Management" permits heads of federal and DoD agencies to exempt certain military tactical, law enforcement, emergency vehicles, and other vehicle classes or types from the requirement to use biofuels.
[45] Components refer to military installations with missions under *U.S. Code* Title 10.
[46] DOE, "Energy and Defense Departments Announce Agreement to Enhance Cooperation on Clean Energy and Strengthen Energy Security," *News*, www.energy.gov/news/9278.htm.
[47] DLA Energy, *Alternative Fuels, Renewable Energy & Energy Conservation* Brochure.
[48] See footnote 9, this chapter.
[49] Jim Eberhardt, DLA Energy, Personal Communication, 2010.
[50] Kelly Widener, "Defense Department, Airlines Sign Alternative Fuels Pact," Defense Energy Support Center, March 22, 2010.
[51] DoD, DoD Instruction 4170.11, "Installation Energy Management," December 11, 2009, army-energy.hqda.pentagon.mil/policies/4170_11.asp.
[52] Army Senior Energy Council and the Office of the Deputy Assistant Secretary of the Army for Energy and Partnerships, *Army Energy Security Implementation Strategy*, January 13, 2009, www.asaie.army.mil/Public/Partnerships/doc/AESIS_13JAN09_Approved%204-03-09.pdf.
[53] RDECOM, "About CERDEC," September 10, 2010, www.cerdec.army.mil/about/index.asp.
[54] RDECOM, "About Us," *Aviation & Missile Research Development & Engineering Center*, www.redstone.army.mil/amrdec/About/index.html.
[55] Kari Hawkins, U.S. Army Garrison Redstone, "Flying Skies With Coal Mix Fuel," *News Front Page*, www.army.mil/-news/2010/07/30/43075-flying-skies-with-coal-mix-fuel/.
[56] Army Energy Program, army-energy.hqda.pentagon.mil/programs/afv1.asp.
[57] See footnote 17, this chapter.
[58] Assistant Secretary of the Air Force for Installations, Environment and Logistics, *Air Force Energy Plan 2010*, www.safie.hq.af.mil/shared/media/document/AFD-091208-027.pdf.
[59] U.S. Air Force, "Alternative Fuels," www.safie.hq.af.mil/shared/media/document/AFD- 080514-067.pdf.
[60] Personal communication with the Naval Supply and Systems Command (NAVSUP) Alter- native Fuels Logistics Program Manager, October 6, 2010.
[61] Liz Wright, Navy Office of Information, "Navy Tests Biofuel-Powered 'Green Hornet,'" April 22, 2010, www.navy.mil/search/display.asp?story_id=52768.
[62] U.S. Marine Corps Energy Summit, August 13, 2009, Washington, DC.
[63] Assistant Commandant of the Marine Corp, Memorandum 11/09, "Establishment of the Marine Corps Expeditionary Energy Office," November 19, 2009.
[64] EIA, *Annual Energy Outlook 2010*, DOE/EIA-0383(2010), May 11, 2010.
[65] Renewable Fuels Association, *2010 Ethanol Industry Outlook*, February 2010, www.ethanolrfa.org/pages/annual-industry-outlook.
[66] See footnote 2, this chapter.
[67] Government Accountability Office (GAO), *Biofuels: DOE Lacks a Strategic Approach to Coordinate Increasing Production with Infrastructure Development and Vehicle Needs*, GAO-07- 713, June 2007.
[68] See footnote 1, this chapter.
[69] Don Hofstrand, Iowa State University, "Tracking Ethanol Profitability," *Ag Decision Maker*, July 2010.
[70] USDA, *Agricultural Projections to 2019*, 2010.
[71] USDA, Crop Production 2000 Summary, usda.mannlib.cornell.edu/usda/nass/CropProdSu//2000s/2000/CropProdSu-01-12-2000.pdf.
[72] See footnote 7, this chapter.
[73] Federal Register, *EPA: Regulation of Fuels and Fuel Additives, Changes to Renewable Fuel Standard Program: Final Rule*, 75 (58): 14670–14904, March 26, 2010.
[74] Aventine Renewable Energy, Inc., presentation at OPIS Energy and Biodiesel Summit, March 1, 2007.
[75] Goldman Sachs Group, *OPIS Ethanol and Biodiesel Supply Summit: Wall Street View of Ethanol Sector*, March 1, 2007.
[76] CBO, *The Impact of Ethanol Use on Food Prices and Greenhouse-Gas Emissions*, 2009.
[77] See footnote 10, this chapter.
[78] USDA Economic Research Service, *Sugar and Sweeteners: Data Tables*, July 2010, www.ers.usda.gov/Briefing/Sugar/Data.htm.
[79] USDA, *The Economic Feasibility of Ethanol Production from Sugar in the United States*, July 2006.
[80] See footnote 10, this chapter.

[81] DOE and USDA, *Biomass as feedstock for a bioenergy and bioproducts industry: The technical feasibility of a billion-ton annual supply*, 2005.

[82] DOE EIA, *Energy and economic impacts of implementing both a 25-percent renewable portfolio standard and a 25-percent renewable fuel standard by 2025*, 2007.

[83] Union of Concerned Scientists, *Climate 2030: A National Blueprint for a Clean Energy Economy,* 2009.

[84] Estimated specifically for cellulosic ethanol. J. Richard Hess, Christopher T Wright, and Kevin L Kenney, "Cellulosic biomass feedstocks and logistics for ethanol production," Idaho National Laboratory, Wiley InterScience, *Biofuels, Bioprod. Bioref* (2007), 1:181–190.

[85] Renewable Fuels Association, *2010 Ethanol Industry Outlook,* February 2010, www.ethanolrfa.org/pages/annual-industry-outlook.

[86] See footnote 22, this chapter.

[87] See footnote 10, this chapter.

[88] EPA, *Renewable Fuel Standard Program (RFS2) Regulatory Impact Analysis*, EPA-420-R- 10-006, February 2010.

[89] See footnote 6, this chapter.

[90] Renewable Fuels Association, *2010 Ethanol Industry Outlook,* February 2010.

[91] Jessica Lieber, "Economics Improve for First Commercial Cellulosic Ethanol Plants," *New York Times*, February 16, 2010.

[92] John M. Urbanchuk, *Importance of the VEETC to the U.S. Economy and the Ethanol Indus- try*, Renewable Fuels Association, March 18, 2010.

[93] Bruce Peterson, Shih-Miao Chin, Sujit Das, *Analysis of Fuel Ethanol Transportation Activi- ty and Potential Distribution Constraints*, Oak Ridge National Laboratory, 2009.

[94] Patricia Ellis, "Ethanol, Will It Drive You to Drink?" presentation at EPA Region 3 LUST Conference, Delaware Department of Natural Resources and Environmental Control, April 2006.

[95] See footnote 4, this chapter.

[96] See footnote 30, this chapter.

[97] Progressive Railroading, *Fleet Stats 2010: Freight car, locomotive and passenger rail car data,* August 2010.

[98] Ken Columbia, "Trains, Trucks, Tanks & Barges," presentation at National Biodiesel Board Convention, World Energy, February 6, 2007.

[99] DOE, Office of Energy Efficiency and Renewable Energy, *Clean Cities Alternative Fuel Price Report,* March 2007.

[100] See footnote 25, this chapter.

[101] DOE, Alternative Fuels and Advanced Vehicles Data Center, *Alternative Fueling Station Total Counts by State and Fuel Type*, August 2010, www.afdc.energy.gov/afdc/fuels/stations_counts.html.

[102] Logan Caldwell, "The Changing Ethanol Market: Implications for Stakeholders," *Energy Producer Magazine*, July 2007.

[103] See footnote 36, this chapter.

[104] These products can be methyl, propyl, or ethyl esters.

[105] As with conventional petroleum production, cellulosic diesel fuel (and renewable diesel) will be the primary fuel product focus rather than jet fuel, which is considered to be a valuable coproduct fuel. Due to the profit and RFS2 mandate drivers, producers are likely to "tune" produc- tion to maximize cellulosic diesel and gasoline, which will minimize the volumes of cellulosic (and renewable) jet fuels produced (~10 percent of the fuel production fraction).

[106] FAPRI, *U.S. and World Agricultural Outlook*, January 2010, FAPRI Staff Report 10-FSR 1, ISSN 1534-4533, Iowa State University, University of Missouri-Columbia, Ames, IA (FAPRI6US).

[107] See footnote 3, this chapter.

[108] National Biodiesel Board, *U.S. Biodiesel Production Capacity,* June 2009, www.biodiesel.org/pdf_files/fuelfactsheets/Production_Capacity.pdf.

[109] See footnote 5, this chapter.

[110] See footnote 5, this chapter.

[111] EIA, AEO 2007, DOE/EIA-0383 (2007), February 2007.

[112] ASTM D975 permits conventional diesel fuels to contain up to 5 percent (including B2 and B5 blends). Commercial biodiesel blends can contain between 6 and 20 percent biodiesel per ASTM D7467. Per current Federal Fleet Guidance (EO 13514 Section 12), blends of B20 must be used in NTVs to count as biodiesel.

[113] National Biodiesel Board (NBB), Biodiesel: Feedstock Supply, *Achieving 1 Billion Gallons of Biodiesel While Protecting Valuable Feedstocks*, www.biodiesel.org/resources/sustainability/pdfs/Achieving%201%20billion %20gallons%20of%2 0fuel%20While%20Protecting%20Valuable%20Feedstocks%20_June%209_.pdf.

[114] USDA, *USDA Agricultural Projections to 2019*, Long-term Projections Report OCE-2010- 1, Interagency Agricultural Projections Committee, February 2010, p. 4.

[115] See footnote 3, this chapter.

[116] EPA, *Renewable Fuel Standard Program (RFS2) Regulatory Impact Analysis*, EPA-420-R- 10-006, February 2010.

[117] See footnote 13, this chapter.
[118] See footnote 13, this chapter.
[119] Rudy Pruszko, "Biodiesel Basics—How It Works & What It Costs," PowerPoint presenta- tion, Center for Industrial Research and Service, Iowa State University Extension.
[120] USDA, *Agricultural Projections to 2019*, 2010.
[121] USDA, National Statistics for Soybeans, quickstats.nass.usda.gov/.
[122] See footnote 17, this chapter.
[123] National Sustainable Agriculture Information Service, *Biodiesel: The Sustainability Dimen- sions*, attra.ncat.org/attra-pub/PDF/biodiesel_sustainable.pdf.
[124] Sam Lines, *An Exploding Market? Utilizing Waste Glycerol from the Biodiesel Production Process*, April 19, 2009, School of Natural Resources and Environment at the University of Michigan.snrecmitigation. wordpress.com/2009/04/19/an-exploding-market-utilizing-waste-glycerol-from-the-biodiesel-production-process/.
[125] See footnote 11, this chapter.
[126] See footnote 13, this chapter.
[127] See footnote 13, this chapter.
[128] John Sheehan, Terri Dunahay, John Benemann, Paul Roessler, *A Look Back at the U.S. De- partment of Energy's Aquatic Species Program: Biodiesel from Algae*, NREL, July 1998.
[129] National Biodiesel Board, *U.S. Biodiesel Production Capacity,* June 2009, www.biodiesel.org/pdf_files/fuelfactsheets/Production_Capacity.pdf.
[130] See footnote 26, this chapter.
[131] Biomass Research and Development Board, *The Economics of Biomass Feedstocks in the United States: A Review of the Literature*, Occasional Paper No. 1, October 2008.
[132] See footnote 13, this chapter.
[133] ExxonMobil and Synthetic Genomics' research effort, which began in 2009 with a $600 million investment, is not expected to yield market-ready fuel for 5 to 6 years.
[134] See footnote 13, this chapter.
[135] See footnote 6, this chapter.
[136] See footnote 6, Chapter 6.
[137] See footnote 13, this chapter.
[138] The cost estimates presented is indicative for a specific renewable diesel pathway using yel- low grease feedstock. Given the growth limits of this feedstock and likely use of multiple oil feedstock for renewable diesel production, the estimate provided here is not identical or appropri- ate for the macro-level costs developed for and presented in Chapter 8.
[139] See footnote 13, this chapter.
[140] DOE, *Biodiesel Handling and Use Guidelines*, DOE/GO-1-2006-2358, EERE, September 2006.
[141] See footnote 36, this chapter.
[142] Nicholas Zeman, "From the plant to the pump," *Biodiesel Magazine*, 2007.
[143] See footnote 36, this chapter.
[144] DOE, Alternative Fuels and Advanced Vehicles Data Center, *Alternative Fueling Station Total Counts by State and Fuel Type*, August 2010, www.afdc.energy.gov/afdc/fuels/ stations_counts.html.
[145] DOE, Energy Efficiency & Renewable Energy, *Clean Cities Alternative Fuel Price Report*, Clean Cities Program, January 2010.
[146] See footnote 41, this chapter.
[147] The EPA estimates that 1,284 mgpy (30 percent) of this 4,280 mgpy total will be produced using Fischer-Tropsch BTL processes, but they express uncertainty regarding the processes will be used to produce the remaining 70 percent of the projected supply. Fuels derived from Fischer- Tropsch BTL processes will likely be more easily be qualified as a drop-in for tactical systems because these fuels are similar to the tested and certified Fischer-Tropsch CTL fuels. In Chapter 8, we assume EPA's 30 percent estimate of fuels anticipated to be derived from Fischer-Tropsch BTL processes would represent the projected supply of drop-in cellulosic diesel that will be available for DoD use in tactical systems.
[148] As of November 2010, the ASTM addendum petition for HRJ is still pending, but technical participants anticipated that it will be approved and released in 2011.
[149] SPK terminology is used for both HRJ and cellulosic jet fuels, which are considered renew- able fuels. However, these fuels should not be confused with alternative fuel CTL or GTL FT-SPK products as these are not renewable fuels.
[150] msuextension.org/publications/AgandNaturalResources/MT200701AG.pdf.
[151] Biomass Advisors. 2010. Camelina Aviation Biofuel Report.
[152] Given USDA does not yet consider camelina a commodity crop, Montana is the only state where some limited production and price information is available.
[153] Personal communication from John Heimlich, Chief Economist, Air Transport Association, September 15, 2010.

[154] This is true of most FT-SPK, whether made from renewable or non-renewable sources. However, older, higher-temperature Fischer-Tropsch technologies (such as Sasol Fischer-Tropsch fuel produced in South Africa) can exhibit lower than average cetane levels, which are damaging to compression ignition engines.

[155] Honorable Ray Mabus, Secretary of the Navy, remarks before the Clean Energy Economy Forum, Washington, DC, July 27, 2010.

[156] DoD Quadrennial Defense Review, 2010.

[157] See footnote 3, this chapter.

[158] DoD, *Strategic Sustainability Performance Plan*, August 26, 2010.

[159] Grace V. Jean, "Navy's Energy Reform Initiatives Raise Concern Among Ship Builders," *National Defense*, April 2010,www.nationaldefensemagazine.org/archive/2010/April/Pages/NavyEnergyReformRaiseConcerns.aspx.

[160] This assumes that renewable fuel prices are not tied to petroleum prices.

[161] See footnote 5, this chapter.

[162] EPA, *EPA Finalizes Regulations for the National Renewable Fuel Standard Program for 2010 and Beyond*, Report EPA-420-F-10-007, February 2010.

[163] Commercial viability assumes capital funding, design, and construction resources will be available for new cellulosic biodiesel plant starts.

[164] DOE, Advanced Fuels & Advanced Vehicles Data Center, www.afdc.energy.gov/afdc/laws/matrix/tech.

[165] DOE (2008) World Biofuels Production Potential.

[166] Official Journal of the European Union (2009), Directive 2008/101/EC of the European Parliament and of the Council of November 19, 2008, amending Directive 2003/87/EC so as to include aviation activities in the scheme for GHG emission allowance trading within the Community, L 8/3-8/21, eurlex.europa.eu/LexUriServ/LexUriServ.do?uri=OJ:L:2009:008: 0003:0021:EN:PDF.

[167] Per ASTM D975 permits conventional diesel fuels to contain up to 5 percent (i.e., including B2 and B5 blends). Commercial biodiesel blends can contain between 6–20 percent biodiesel per ASTM D7467. Per current Federal Fleet Guidance (EO 13514 Section 12), blends of B20 must be used in NTVs in order to count as biodiesel.

[168] DoD has established a process to evaluate candidate fuels (MIL-HDBK-510). The Air Force and Navy are certifying aircraft fuels, while the Army is evaluating ground vehicle fuels.

[169] Table 4.1, Defense Science Board Task Force on DoD Energy Strategy, *More Fight Less Fuel,* February 2008, handle.dtic.mil/100.2/ADA477619.

[170] Personal communication with Air Force Service Control Point and Fuels representatives, May 21, 2010.

[171] Personal communication with the NAVSUP Alternative Fuels Logistics Program Manager, October 6, 2010.

[172] Figures 8-1 and 8-2 projections for renewable cellulosic diesel supply are based on EPA's estimate that 30 percent of cellulosic diesel fuel produced in the United States will use Fischer-Tropsch BTL processes. This differs from Figure 6-1 forecasts of the overall U.S. cellulosic diesel projected total, which includes drop-in renewable fuels (estimated in Figures 8-1 and 8-2) and fuel from other processes, which could require additional testing and certification prior to being a drop- in cellulosic diesel.

[173] In FY09, DoD fleet vehicles consumed 2.7 million GGEs of E85 and 5.1 million GGEs of biodiesel.

[174] Personal communication with John Heimlich, Vice President and Chief Economist, Air Transport Association, September 15, 2010. The lower estimates assume that federal tax credits and other incentives will be in place.

[175] Our cost projections are relative to today's cost of petroleum products projected forward from 2015 through 2020. Should the cost of crude oil significantly rise, it is likely that this cost difference would close if not eliminate the price premium. However, production of renewable fuels requires energy and their costs would be anticipated to rise as well. Agricultural commodity prices could possibly rise more than crude oil costs within the next few years. Nevertheless, investment in renewable fuels could help to provide a hedge against a sudden and steep rise in petroleum prices if these fuels were being obtained under long-term contracts at fixed prices.

[176] ASTM has formed a task force to examine alternative diesel fuels' "fit for purpose" proper- ties and confirm that they can actually be considered "drop-in" fuel compatible with existing infra- structure.

[177] Personal communication from Lynda M. Turner, November 19, 2010.

[178] See footnote 24, this chapter.

[179] See footnote 16, this chapter.

[180] Fischer-Tropsch fuels can be produced from renewable (cellulosic) and non-renewable (coal and natural gas) feedstocks. However, it is not outside the realm of technical feasibility to utilize both renewable and non-renewable feedstocks using a hybrid process. Given the environmental and cost-effectiveness tradeoffs, it is likely that these hybrid production plants might arise, but it remains an open question whether they would be considered a renewable fuel under RFS2.

[181] Chevron, *Alternative Jet Fuels: A supplement to Chevron's Aviation Fuels Technical Review*, www.chevronglobalaviation.com/docs/5719_Aviation_Addendum._webpdf.pdf.

[182] NASA, "Alternate Fuels for Use in Commercial Aircraft," 18th ISABE Conference, NASA/TM-2008-214833, 2008, gltrs.grc.nasa.gov/reports/2008/TM-2008-214833.pdf.

[183] See footnote 17, this chapter.

[184] Personal communication from Jeffrey J. Braun, September 10, 2010.

[185] Joint Staff, Joint Publication 4-09, Chapter 4, "Joint Doctrine For Global Distribution," February 5, 2010, www.dtic.mil/doctrine/jel/new_pubs/jp4_09b.pdf.

[186] See www.desc.dla.mil/DCM/DCMPage.asp?PageID=66.

[187] An NSN is an official label applied to items repeatedly bought, stocked, stored, issued, and used throughout the federal supply system.

[188] EPA, *EPA Finalizes Regulations for the National Renewable Fuel Standard Program for 2010 and Beyond*, Report EPA-420-F-10-007, February 2010.

[189] Commercial viability assumes capital funding, design, and construction resources will be available for new cellulosic biodiesel plant starts.

[190] DOE, Advanced Fuels & Advanced Vehicles Data Center, www.afdc.energy.gov/afdc/laws/matrix/tech.

[191] DOE (2008) World Biofuels Production Potential.

[192] See footnote 4, this appendix.

[193] Official Journal of the European Union (2009), Directive 2008/101/EC of the European Parliament and of the Council of November 19, 2008, amending Directive 2003/87/EC so as to in- clude aviation activities in the scheme for GHG emission allowance trading within the Community, L 8/3-8/21, eur-lex.europa.eu/ LexUriServ/LexUriServ.do?uri=OJ:L:2009:008: 0003:0021:EN:PDF.

[194] Per ASTM D975 permits conventional diesel fuels to contain up to 5 percent (i.e., including B2 and B5 blends). Commercial biodiesel blends can contain between 6–20 percent biodiesel per ASTM D7467. Per current Federal Fleet Guidance (EO 13514 Section 12), blends of B20 must be used in NTVs in order to count as biodiesel.

[195] EIA, *Annual Energy Outlook 2010*, DOE/EIA-0383(2010), May 11, 2010.

[196] Renewable Fuels Association, *2010 Ethanol Industry Outlook*, February 2010, www.ethanolrfa.org/pages/ annual-industry-outlook.

[197] See footnote 2, this appendix.

[198] GAO, *Biofuels: DOE Lacks a Strategic Approach to Coordinate Increasing Production with Infrastructure Development and Vehicle Needs*, GAO-07-713, June 2007.

[199] See footnote 1, this appendix.

[200] Don Hofstrand, Iowa State University, "Tracking Ethanol Profitability," *Ag Decision Maker*, July 2010.

[201] USDA, *Agricultural Projections to 2019*, 2010.

[202] USDA, Crop Production 2000 Summary, usda.mannlib.cornell.edu/usda/nass/CropProdSu//2000s/2000/ CropProdSu-01-12-2000.pdf.

[203] See footnote 7, this appendix.

[204] See footnote 7, this appendix.

[205] Aventine Renewable Energy Inc., *Presentation at OPIS Energy and Biodiesel Summit*, March 1, 2007.

[206] Goldman Sachs Group, *OPIS Ethanol and Biodiesel Supply Summit: Wall Street View of Ethanol Sector*, March 1, 2007.

[207] National Corn Growers Association, *How much ethanol can come from corn?* February 2007.

[208] CBO, *The Impact of Ethanol Use on Food Prices and Greenhouse-Gas Emissions*, 2009.

[209] *Federal Register*, "EPA: Regulation of Fuels and Fuel Additives, Changes to Renewable Fuel Standard Program: Final Rule," 75 (58): 14670–14904, March 26, 2010.

[210] USDA Economic Research Service, *Sugar and Sweeteners: Data Tables*, July 2010, www.ers.usda.gov/ Briefing/Sugar/Data.htm.

[211] USDA, *The Economic Feasibility of Ethanol Production from Sugar in the United States,* July 2006.

[212] See footnote 15, this appendix.

[213] DOE and USDA, *Biomass as feedstock for a bioenergy and bioproducts industry: The technical feasibility of a billion-ton annual supply*, 2005.

[214] See footnote 19, this appendix.

[215] DOE EIA, *Energy and economic impacts of implementing both a 25-percent renewable portfolio standard and a 25-percent renewable fuel standard by 2025*, 2007.

[216] Union of Concerned Scientists, *Climate 2030: A National Blueprint for a Clean Energy Economy,* 2009.

[217] Estimated specifically for cellulosic ethanol. J. Richard Hess, Christopher T. Wright, and Kevin L. Kenney, "Cellulosic biomass feedstocks and logistics for ethanol production," Idaho National Laboratory, Wiley InterScience, *Biofuels, Bioprod. Bioref,* 1:181-190 (2007).

[218] DOE, *Biomass Multi-Year Program Plan*, Office of the Biomass Program, Energy Efficiency and Renewable Energy, 2010, p. 3–21.

[219] William T. Coyle, USDA, *Next-Generation Biofuels: Near-Term Challenges and Implications for Agriculture*, BIO-01-01, May 2010, p. 11, citing Hudson, 2009.

[220] Congressional Research Service, *Cellulosic Biofuels: Analysis of Policy Issues*, February 1, 2010, p. 12.

[221] See footnote 2, this appendix.

[222] See footnote 2, this appendix.

[223] See footnote 8, this appendix.

[224] EPA, *Renewable Fuel Standard Program (RFS2) Regulatory Impact Analysis*, EPA-420-R-10-006, February 2010.
[225] See footnote 6, this appendix.
[226] See footnote 2, this appendix.
[227] Jessica Lieber, "Economics Improve for First Commercial Cellulosic Ethanol Plants," *New York Times*, February 16, 2010.
[228] See footnote 25, this appendix.
[229] See footnote 33, this appendix.
[230] John M. Urbanchuk, *Importance of the VEETC to the U.S. Economy and the Ethanol Industry*, Renewable Fuels Association, March 18, 2010.
[231] Bruce Peterson, Shih-Miao Chin, Sujit Das, *Analysis of Fuel Ethanol Transportation Activity and Potential Distribution Constraints*, Oak Ridge National Laboratory, 2009.
[232] See footnote 15, this appendix.
[233] See footnote 37, this appendix.
[234] See footnote 4, this appendix.
[235] Progressive Railroading, *Fleet Stats 2010: Freight car, locomotive and passenger rail car data*, August 2010.
[236] Ken Columbia, World Energy, "Trains, Trucks, Tanks & Barges," presentation at National Biodiesel Board Convention, February 6, 2007.
[237] John Whims, *Pipeline Considerations for Ethanol,* August 2002.
[238] NREL, *Biofuels Availability and Use*, June 2007.
[239] American Commercial Lines, "Moving America 2007," presentation at OPIS Ethanol and Biodiesel Summit, March 2007.
[240] See footnote 30, this appendix.
[241] See footnote 42, this appendix.
[242] Brat, Ilan, and Daniel Machalaba, "Can Ethanol Get a Ticket to Ride?" *The Wall Street Journal*, February 1, 2007.
[243] See footnote 32, this appendix.
[244] See footnote 30, this appendix.
[245] Patricia Ellis, "Ethanol, Will It Drive You to Drink?" presentation at EPA Region 3 LUST Conference, Delaware Department of Natural Resources and Environmental Control, April 2006.
[246] See footnote 30, this appendix.
[247] DOE, Alternative Fuels and Advanced Vehicles Data Center, *Alternative Fueling Station Total Counts by State and Fuel Type*, August 2010, www.afdc.energy.gov/afdc/fuels/stations_counts.html.
[248] Logan Caldwell, "The Changing Ethanol Market: Implications for Stakeholders," *Energy Producer Magazine*, July 2007.
[249] DOE, Office of Energy Efficiency and Renewable Energy, *Clean Cities Alternative Fuel Price Report*, March 2007.
[250] These can be either methyl, propyl, or ethyl esters.
[251] As with conventional petroleum production, cellulosic diesel fuel (and renewable diesel) will be the primary fuel product focus rather than jet fuel, which is considered to be a valuable coproduct fuel. Due to the profit and RFS2 mandate drivers, it is likely that producers will "tune" production to maximize cellulosic diesel and gasoline, which will minimize the volumes of cellulosic (and renewable) jet fuels produced (i.e., ~10% of the fuel production fraction).
[252] FAPRI, *U.S. and World Agricultural Outlook*, January 2010, FAPRI Staff Report 10-FSR 1, ISSN 1534-4533, Iowa State University, University of Missouri-Columbia, Ames, IA (FAPRI6US).
[253] See footnote 3, this appendix.
[254] National Biodiesel Board, *U.S. Biodiesel Production Capacity,* June 2009, www.biodiesel.org/pdf_files/fuelfactsheets/Production_Capacity.pdf.
[255] National Biodiesel Board (NBB), Biodiesel: Feedstock Supply, *Achieving 1 Billion Gallons of Biodiesel While Protecting Valuable Feedstocks*, www.biodiesel.org/resources/sustainability/pdfs/Achieving%201%20billion% 20gallons%20of%20fuel%20While%20Protecting%20Valuable%20Feedstocks%20_June%209_.pdf
[256] USDA, *USDA Agricultural Projections to 2019*, Interagency Agricultural Projections Committee, February 2010, Long-term Projections Report OCE-2010-1, p. 4.
[257] See footnote 3, this appendix.
[258] See footnote 6, this appendix.
[259] See footnote 7, this appendix.
[260] See footnote 3, this appendix.
[261] See footnote 6, this appendix.
[262] See footnote 6, this appendix.
[263] See footnote 6, this appendix.
[264] Rudy Pruszko, "Biodiesel Basics—How it Works & What it Costs," PowerPoint presentation, Center for Industrial Research and Service, Iowa State University Extension.

[265] USDA, Agricultural Projections to 2019, 2010.
[266] USDA, National Statistics for Soybeans, quickstats.nass.usda.gov/.
[267] See footnote 13, this appendix.
[268] National Sustainable Agriculture Information Service, attra.ncat.org/attra-pub/ PDF/biodiesel_sustainable.pdf.
[269] Sam Lines, An Exploding Market? Utilizing Waste Glycerol from the Biodiesel Production *Process*, April 19, 2009, School of Natural Resources and Environment at the University of Michigan. snrecmitigation. wordpress.com/2009/04/19/an-exploding-market-utilizing-waste-glycerol-from-the-biodiesel-production-process/.
[270] See footnote 6, this appendix.
[271] See footnote 6, this appendix.
[272] See footnote 6, this appendix.
[273] John Sheehan, Terri Dunahay, John Benemann, Paul Roessler, A Look Back at the U.S. Department of Energy's Aquatic Species Program: Biodiesel from Algae, NREL, July 1998.
[274] DOE Biomass Program, National Algal Biofuels Technology Roadmap, 2010.
[275] DOE, *EIA Monthly Energy Review*, www.eia.doe.gov/emeu/mer/environ.html.
[276] See footnote 25, this appendix.
[277] See footnote 26, this appendix.
[278] See footnote 26, this appendix.
[279] See footnote 5, this appendix.
[280] See footnote 5, this appendix.
[281] Biomass Research and Development Board, *The Economics of Biomass Feedstocks in the United States: A Review of the Literature*, Occasional Paper No. 1, October 2008.
[282] See footnote 6, this appendix.
[283] See footnote 6, this appendix.
[284] ExxonMobil and Synthetic Genomics' research effort, which began in 2009 with a $600 million investment, is not expected to yield market-ready fuel for 5 to 6 years.
[285] See footnote 6, this appendix.
[286] Peter J. le B. Williams and Lieve M. L. Laurens, "Microalgae as biodiesel & biomass feedstocks: Review & analysis of the biochemistry, energetics, & economics," *Energy & Environmental Science*, February 2010.
[287] See footnote 37, this appendix.
[288] Abayomi O. Alabi; Martin Tampier; and Eric Bibeau, Microalgae Technologies & Processes for Biofuels/ Bioenergy Production in British Columbia: Current Technology, Suitability, & Barriers to *Implementation*, British Columbia Innovation Council, January 2009.
[289] See footnote 25, this appendix.
[290] Peer M. Schenk et al., "Second Generation Biofuels: High-Efficiency Microalgae for Biodiesel Production," *BioEnergy Research*, March 2008.
[291] www.oilgae.com/algae/oil/extract/che/che.html.
[292] www.mybiofuels.net/MA_Extraction.htm#4.
[293] See footnote 43, this appendix.
[294] Don Hofstrand, Iowa State University, *Tracking Biodiesel Profitability*, Ag Decision Maker, August 2010.
[295] See footnote 45, this appendix.
[296] See footnote 6, this appendix.
[297] See footnote 6, this appendix.
[298] DOE, *Biodiesel Handling and Use Guidelines*, DOE/GO-1-2006-2358, EERE, September 2006.
[299] See footnote 49, this appendix.
[300] Nicholas Zeman, "From the plant to the pump," *Biodiesel Magazine*, 2007.
[301] See footnote 51, this appendix.
[302] See footnote 51, this appendix
[303] DOE, Alternative Fuels and Advanced Vehicles Data Center, *Alternative Fueling Station Total Counts by State and Fuel Type*, August 2010, www.afdc.energy.gov/afdc/fuels/stations_counts.html.
[304] DOE, Energy Efficiency & Renewable Energy (EERE), *Clean Cities Alternative Fuel Price Report*, Clean Cities Program, January 2010.
[305] See footnote 55, this appendix.
[306] Butamax® Biobutanol Fact Sheet, 2009: www.butamax.com/_assets/pdf/biobutanol_a_more_advanced_biofuel. pdf.
[307] DOE EERE Alternative Fuels and Advanced Vehicles Data Center.
[308] Next-Generation Biofuels: Near-Term Challenges and Implications for Agriculture, by Wil-liam Coyle, BIO-01-01, USDA, Economic Research Service, May 2010.
[309] Department of Energy Alternative Fuels & Advanced Vehicles Data Center, 2010.
[310] Biomass Advisors. 2010. Camelina Aviation Biofuel Report.
[311] Source: 40 CFR 80.1401.

INDEX

C

D

E

F

G

H

I

J

K

L

M

N